Ken Spencer of Burnley
Ornithologist and Historian

by
Stephen Ward

Copyright © 2021 Stephen Ward

All Rights Reserved

ISBN: 978-1-911138-33-4

Cover: Ken with pigeon, 12th March 1959 (*Burnley Express*).

Published by Nu-Age Print & Copy
289 Padiham Road, Burnley
BB12 0HA
01282 413373

This biography is dedicated to:

- **Philip Spencer** March 1922 – November 1943;
- **Barbara Bailey** February 1922 - July 2004, Ken's beloved partner and one-time Head of Heasandford Primary School;
- **Susan Halstead & Diane Schofield** Burnley Reference Librarians until 2009;
- **Julie Dawson** Burnley Library 2016, who made arrangements for Ken's MBE;
- **Karen Strachan** who cleaned for Ken & Barbara for 25 years and continued to ensure that Ken got at least one good meal a week after Barbara died, supplemented in recent years by **Susan, Diane, Julie & Tracey**;
- **the late Gerry Almond** - Sunday morning bird & butterfly companion;
- **Jeff & the late Sheila Haworth, Peter Hornby** - bird, botany and butterfly recorders;
- his many **friends at the Weavers' Triangle**;
- **David Smith and colleagues** at St Peter's Parish Church, organisers of the Saturday morning concerts and Thursday lunches;
- **Molly Haines and colleagues** at the Burnley Historical Society
- **David Bridge & Brian Howe**, neighbours on either side, who maintained the fabric of 167, preventing it becoming, in Ken's words, 'like Clitheroe Castle';
- **Tracey Parker** - wildfowl enthusiast;
- **Karl Munslow & Mick Malone** who made Ken a new garden gate.

CONTENTS

Plates viii

Preface xi

1. Early years 1928-1946: Philip and Ken 1
2. Diary for 1941 10
3. Down the Pit 20
4. Ornithologist 22
5. Birds in Poetry 41

   Plates 1-31 45

6. Animal Welfare 62
7. Historian 68
8. The By-Laws Dispute 79
9. Foot & Mouth Disease 84
10. Photochemical Smog 88
11. Civic Pride 92
12. Enthusing Others 97
13. Star Girl at Evening 100
14. Days Out 104
15. Butterfly Days 109
16. On Life 112
17. Recognition 118

    Annex: BTO Gold Pin 126

PLATES

1. Ken at one month and Philip aged six with Mother – July 1928. *(T. Taylor)*.

2. Ken aged two and Philip aged eight with Mother – November 1930. *(T. Taylor)*.

3. Ken aged two and Philip eight with toy plane. – November 1930. *(T. Taylor)*.

4. Ken on 4th June 1930 with Pippin. *(Home photograph)*.

5. Ken on 24th June (his 3rd birthday) 1931 – playing with Philip in field at Lower Holes Farm to rear of Landseer Close. *(Home photograph)*.

6. Philip and Ken about 1938 or 1939 *(probably T. Taylor)*.

7. Philip and Ken about 1940. *(T. Taylor)*.

8. Bus marooned in snow near The Waggoners, 1940. *(Philip Spencer)*.

9. Fields at Lower Holes Farm with Landseer Close in the background, taken by Philip Spencer, 15th April 1940. *(Burnley Reference Library Collection)*

10. As a coal-mining optant in 1947 at the back-door of 18 St. Matthew Street. *(Home photograph)*.

11. Ken on his 21st birthday, 1949. *(Taylor and Lambert, 105 Manchester Road, Burnley)*.

12. Flyer for *The Lapwing in Britain* published in 1953.

13. Taking part in athletics event held in a Clitheroe street about 1955.

14. Graduation Day 28th June 1957, Leeds Town Hall – *(Home photograph)*.

15. Ken's mother – probably on the occasion of his graduation.

16. Ken with pipe – *(Taylor and Lambert)*.

viii

17. Clearing scrub from Meathop Moss with Clive Hartley on a Conservation Corps work party during 1960s *(Home photograph)*.
18. Reclamation at Towneley Colliery spoil heap during the 1960s I. *(Burnley Express)*.
19. Reclamation at Towneley Colliery spoil heap during the 1960s II. *(Burnley Express)*.
20. Planting bulbs with Girl Guides at Holy Trinity Church *(Burnley Express)*.
21. Members of the Civic Trust planting bulbs. Front from left: Ken, Victoria Towneley, Ian Rawson, Peregrine Towneley, Simon Towneley. Back: Barbara, Tom Huggon, J. Myers. *(Burnley Express 14th October 1969)*.
22. With Stanley Jeeves of the Council for the Protection of Rural England and helpers clearing refuse near Briercliffe about 1970.
23. With Tony Audus, Burnley Parks and Recreation Department, and Towneley Junior Naturalists' Club planting snowdrop bulbs supplied by Burnley Civic Trust in Towneley Park. *(Burnley Express and News, 7th November 1972)*.
24. Founders of the Civic Trust: Harry Schofield, Ken and Simon Towneley. *(Burnley Express)*.
25. Civic Trust: Ken with Mary Davenport – one time teacher at the High School. *(Burnley Express)*.
26. Ken with Jane about 1979 at Healey Heights. *(Burnley Express)*.
27. Ken with party conducting judge through Towneley re Burnley by-laws dispute about 1978.
28. Ken relaxing at 167 Manchester Road, undated *(photograph by Barbara)*.
29. Barbara – shortly before she retired *(School photograph)*.

30. Ken with Lawrence Chew at the Weavers' Triangle visitor centre in April 2007. *(Roger Creegan)*.

31. The author with Ken in Scott Park, Christmas Day 2009. *(Sharon Parr)*.

PREFACE

As a boy, I lived at a house called The Nook in Simonstone Lane. Further down the Lane, at Allandale, lived my 'best mate' – Robin Burr. Robin and I used to explore, by bike and on foot, the lanes, fields and woods beyond Simonstone and Altham up Huncoat Lane as far as the Leeds and Liverpool Canal. We became interested in identifying the birds which we saw and that led us, on the 7th April 1958, to join the recently formed East Lancashire Ornithologists' Club which met monthly in Burnley. At the end of each meeting, members recounted local records. Robin and I reported that we had seen a green sandpiper at a 'tarn' just below the canal. This raised a few eyebrows, not least from the club secretary, Len Bouldin, who assigned to Ken the task of checking whether these lads knew what they were talking about.

That was in 1959. We led Ken around our haunts; at first the green sandpiper proved elusive but it led to a series of letters from Ken which was to run for more than 50 years. The first of those letters dates from a Friday night in February 1959; it says *"As you will know, I have been out in the Altham area today in search of the green sandpiper. I checked both tarns, both sewage [works] and the Calder as far as the sewage. No luck."* But on 8th March came the vindication Robin and I sought. *"I paid an unforeseen visit to the tarns this morning and, you will be glad to know, flushed the green sandpiper ... I must thank you both for having given me the word about the bird's presence in the area."*

I wish I could say that I had kept the myriad of letters I received from Ken over the years but, regrettably, it was 1987 before it began to dawn on me that these might be an important record in their own right. Now I have 'archived' them with all the respect they deserve, each one – often with numerous enclosures – housed in its own protective transparent pocket and kept in a ring-backed file, of which there are several. It is my intention that these be officially archived at the Lancashire Record Office.

Ken devoted his whole life to his home town of Burnley. When he reached his late 70s he suffered the loss of his life-time partner, Barbara Bailey, and increasingly adopted Burnley Reference Library as his second home where, having trained as a librarian many years previously and, being steeped in local history, he was able to provide a much-valued service assisting the library staff in answering queries relating to the area. Ken has written a number of books, of which more anon. In 2006, or thereabouts, I suggested to him that he write his autobiography. I was the second person to suggest this in as many weeks, the first being Susan Halstead, at that time the Reference Librarian in Burnley Central Library. When, after two years, Ken reached his 80th birthday and it became increasingly obvious that this was one book he was not inclined to write, I told him I had no other option but to take on the task myself and write his biography.

What follows is the story of a remarkable man:

- passionate about poetry – especially that of his elder brother Philip who was tragically killed in the Second World War at the age of only 21;
- an expert on poetry featuring birds and, indeed, in his own quiet way, a poet himself;
- an ornithologist who wrote what was, for more than half-a-century, the standard work on the lapwing and who, Ken told me more than once, would have liked to become a professional in that sphere;
- a butterfly expert from boyhood to which he returned in later life;
- a one-time special constable for the Bird Protection Act 1954;
- a nature conservationist who led young parties both locally and on behalf of the then Conservation Corps [later to become The Conservation Volunteers] undertaking management work on sites as varied as colliery spoil-heaps, National Nature Reserves, and those of the Wildlife Trusts and RSPB;

- a supporter for more than 60 years of the Royal Society for the Protection of Birds, the British Trust for Ornithology and the British Ornithologists' Union;
- a teacher of English to students at the technical colleges in both Accrington and Burnley;
- a caring citizen and one of the founding members and one-time chairman of the Burnley and District Civic Trust which has done so much to retain features of local and national importance;
- a local historian and one-time chairman of the Burnley and District Historical Society;
- a highly principled man who rose to prominence during the Burnley by-laws dispute, but who has campaigned tirelessly over the years on planning matters from open-cast coal-mining to the siting of windfarms and,
- in the light of Barbara's painful and lingering death from cancer, a campaigner in the cause of assisted dying.

Stephen Ward
Herons Reach
Arnside

December 2020

CHAPTER 1
Early years 1928 to 1946: Philip and Ken
Plates 1-7 and 9

Ken Spencer was born on the 24th of June 1928 to Cyril Bickersteth Spencer, a cotton manufacturer, and Maria Ermintrude Spencer. Ken is an abbreviation of Kenrick - a character in *The Thirsty Sword: a story of the Norse invasion of Scotland 1262-63*. The family lived at Oak Dene, Landseer Close in Burnley. To the front the house overlooks Scott Park; to the rear, where it is now hemmed in by the town, lay farmland. Oak Dene gave Ken and his brother Philip, six years his senior, a base from which to play in open fields. On the far side of the park lies Holly Mount Terrace where their mother's sister, Aunt Gert, lived at 18 St. Matthew Street which was home from home.

In 1933 Ken was a pupil at Sunny Bank School, 171 Manchester Road, locally known as 'Bunny Swank'. He enjoyed Miss Whitehead's teaching of English so much that when he left there in 1939, aged 11, he recalled *"I could have passed the School Certificate in that subject, I feel sure: we worked so well together."* When he became a teacher himself, he used many of her illustrations for his own classes. Long afterwards, he wrote to thank her and was delighted to hear that she recalled both Philip and him with affection and happiness.

Ken sat the 'eleven-plus' in 1939 and went on to Burnley Grammar School. New pupils were able to choose which 'house' they joined, Brun, Calder, Pendle or Ribblesdale (Spencer, 2003). Ken's pal, Richard Chandler, one year his senior, had emphasised he must choose Pendle but by default he ended up in Brun.

Ken would have liked to have taken art but, except where it was essential for a career such as architecture, there was a prejudice against it. On the other hand, to his dismay, since he lacked any affinity with it, mathematics was compulsory.

Ken valued his time in the 6th form: *"I was enthralled by Keats: 'The Eve of St. Agnes' and the 'Ode to a Nightingale' entered my heart and mind and became a part of me ... When I entered the sixth form it was – to quote the title of a Robert Westall story – 'Falling into Glory'. ... learning was almost effortless, a voyage of discovery. I have never been so happy in my life as I was in the first year sixth ... If you know the Dylan Thomas poem 'Fern Hill' you'll understand something of how it was."* Students were treated as young men; lessons were akin to university tutorials, with a stimulating exchange of ideas.

Ken enjoyed cross-country running and athletics. In this, he enjoyed a friendly rivalry with Albert Ashworth. One day, jogging alongside Albert between Heasandford and Queen Victoria Road, they were overtaken by a flat-bed lorry travelling slowly. *"Jump on"*, said Albert, so they did. The difficulty came in alighting so that they could report to the headmaster at Heasandford House. Ken tried to alight at the last bend, but the momentum pitched him forward, sustaining horrific grazes to his left knee and elbow.

Athletics remained an interest and, after the war, he was for many years a member of Clayton-le-Moors Harriers. A 1955 news-cutting reads *"The Clayton-le-Moors Harriers' annual handicap was run last Saturday over 8 miles on the western edge of Pendle, and was won by K.G. Spencer of Burnley."*

Ken remembers the end of the war in Europe: *"We were in a lesson with Mr. Bennett*[1] *; it was early afternoon. A fifth-form boy called Peter Eastwood knocked, came in and quietly said 'I have been sent to tell you that hostilities have ceased.'... Mr. Bennett nodded his thanks and said 'Right, you can all go now' and we quietly dispersed."*[2] Later in the week there were celebrations outside the Town Hall with music and dancing.

---

[1]. Walter Bennett who wrote *The History of Burnley*.
[2]. Spencer, Ken. 2006 *They Made a Difference: Notes on Some Burnley People.* Burnley and District Historical Society p. 6.

The 15 years from Ken's birth in 1928 to Philip's premature death in 1943, had a profound effect on Ken, which lasted his whole life. An inkling of the interaction between the two boys is evident from Philip's poems, some reflecting on their boyhood, some dedicated to Ken, others seeming to anticipate his fate.

The year 1941 is a special one in this account and forms the subject of the next chapter. For that year, and for that year only, at the age of twelve, Ken kept a diary from 1st January to 20th August. In this he may have been inspired by Philip who kept a diary from when he went away to school at the age of 11 in 1933 to the eve of his fatal flying accident in November 1943. Writing in 2007 of his parents' decision to send Philip, but not himself, to boarding school in North Yorkshire, Ken said *"You may wonder why Philip was sent to Richmond School. My Mother – long afterwards – told me why it was. Whilst on a family holiday at Grange, she and my father met a Captain Sanderson, who so enthused over his old school that my parents resolved to send Philip there. Many boys from middle-class families then 'went away to school'. Burnley Grammar School at that time had only an indifferent reputation. I always regretted that Philip did not go there and, in later years, my mother did too. Philip's intervention, she once told me, persuaded my parents to let me go to BGS."*

Looking forward to winter at home, Philip wrote to Ken from St. John's College, Oxford c. 1941, where he read English: *"we're going to get up in the dark some morning – perhaps when there is snow on the ground – and go out and see the dawn break over the hills, and watch the pale sun as it lights the wintry sky. There might be some sheep there, and perhaps an old signpost on Crown Point, and it will be just like a Christmas card."*

In a letter written in July 1942, after he had joined the RAF, Philip asked Ken to tell him all about Mesdames Chalmondley, Ponsonby and Featherstonehaugh. On asking Ken the identity of these three particular ladies, he laughed aloud recalling that they were Rhode Island Red Hens acquired for egg-laying and given these outlandish names by Philip.

Recalling happy days, Philip continued *"I shall often be thinking of Hambledon, the stream at the base of the lesser end, and the single blackthorn tree growing out of the stream, and the curlews, and above all the pond where so many of our boats are still sailing. If ever you go up to the pond, don't forget to set off some boats, will you?"*

DREAM BOATS

1. BOATING

To the Realm of Fancy floating,
  To some fairy land of dreams,
Little paper boats a-boating,
  Sailing down our silver streams.

Over rapids, dipping, spinning,
  Drifting down each slower pool;
On some journey each beginning
  Over water crystal-cool.

Edging into dewy mosses
  Hanging from the verdurous bank,
Catching in the straggling grasses,
  Gliding past the marsh-plants dank.

2. THE RECOLLECTION

Sometimes in a gloomy cavern
  Where the water-boatmen swim
Ships would find themselves a haven,
  Moss-walled, sheltered, cool and dim.

Sometimes in the sunshine glancing
  Wafted by the gentle breeze,
Over ripples lightly dancing,
  They would sail the fairy seas.

And the sound of water, singing,
  As it bubbled on its way.
Was a thousand bluebells ringing
  Breathed on by the breath of May.

3. DREAM BOATS

Dancing through the sylvan sunlight,
  Sailing calmly by the moon,
Riding jewelled neath the starlight, -
  I shall come and find you soon.

Somewhere boats of mine are boating,
  Drifting farther down the streams;
To the Realm of Fancy floating,
  To some fairy land of dreams.

On the day prior to embarkation at Liverpool, bound for Africa, Philip wrote to Ken *"I wrote a very poor sonnet in the Puss Moth Lane a week ago. However, please keep it and in the event of publication of the rest of my stuff, please put it with the manuscripts. This is dedicated to you. I have called it 'The Man at the Gate', or more simply 'Written in Puss Moth Lane, Hoylake, 20th July 1942. Dedicated to Ken'."*

THE MAN AT THE GATE

The evening falls upon my homing days,
Above the unknown night my star is risen.
For the fair day I knew be thanks and praise
To God; and courage, faith to meet the prison

Roomless, viewless, that the future bodes, be mine
From God.
Here where the lane is bitter-sweet
With lights lost from life's multi-toned design, -
While in the present, past and future meet, -

> The last strong link with folks and home I break,
> And sigh for days I know will come no more:
> Here is no change, yet I my farewell take;
> Evening draws on, I turn towards the shore,
>
> The day has closed – like childhood fades the light:
> Now, on, my soul, out, out into the night.

Lancashire remained very much alive for Philip and on 19th March 1943 he dedicated another poem to Ken, which appears in full in the anthology 'African Crocus and other poems':

> Extract from **THE MUSIC OF NAMES**
> (To Ken, because it is about times we shared.)
>
> At the mention of Whitewell there arises a shining magic,
> Cliviger is full of the wine of summer's bowl,
> Hurstwood Lake is hushed with the lull of blue music,
> Sabden, the sorrow, and Newchurch the balm of the soul.
>
> Towneley is dark with thick oaks, and the ash trees' leaves are falling,
> Mention but Pendle and the winds and the curlews would come,
> Speak then of Hambledon – it is like myself that is calling,
> Whisper but Burnley and my soul with my thoughts goes winging home.

In 1943 Ken was awarded 1st prize by the Lancashire and Cheshire Area Economic League (motto *Knowledge by Facts*, for his essay on a lecture *The Making of Today*). This gave rise to a stirring letter from Philip, sent Easter Day 25th April 1943, who by this time was undergoing training as a navigator with the RAF in Rhodesia. Philip had been on leave and spent it locally with friends in Africa. He went on to say: "on my return were dozens of newspapers and letters from home awaiting me. Amongst these papers from home was a report of the meeting of the Lancashire and Cheshire Economic League Youth Movement at which Sir George Paish presided, and at which you received a prize for a literary effort on a political theme and are to be congratulated on this outstanding achievement. This is the main reason that I am writing to you now. Until I had your letter in which you had been to the R. Forbes lecture on Russia I had no idea that you had the political lung in your bed. Now I ask – may I demand – that I hear more about your political views. Am I to believe from the subject of the lecture you attended and the thesis you wrote, that you are Communistic in outlook? Whatever you are, please tell me all about it, and what brought about this way of thinking into your life. Perhaps I shall be able to help you from the small store of experience I have gained from knowledge of Oxford political societies. Anyhow I will give you all the help you might like, incidentally helping myself at the same time. That by the way is a poor thing to apply to politics. Reading Sir George's speech I see that he is imperialistic – 'Britain is best, Britain is the goods' this that and the other to the glory of Britain – but it won't work. We've got to get ourselves in the right perspective – to have a sense of proportion ... I don't know much about Sir George, but I don't like to read of him having said such things as 'Gt. Britain has given greater benefit to the world than any other state in history.' (Including the Romans, I suppose, who gave us our civilisation). And again, as I said at the beginning of this political digression – the incidental process of helping ourselves isn't very commendable. 'We were good neighbours' (said Sir G.) 'and wanted to help everybody, and incidentally helped ourselves in so doing'."

Philip goes on to recount a recent experience. "When I arrived back here from leave it was raining. I was waiting for a taxi beside some Indians on a part of the pavement under shelter. A white man came along, pushed the Indian aside and said 'Think you own the place?' and the Indian murmured something about 'Part of it.' The white man came back and seized the Indian and tried to push him off the pavement but he was not strong enough and he said 'You don't talk to a white man like that – understand?' And he began to argue with the native that he had no business to say that he owned any part of the place, but the Indian was equal to him in argument, and in the end the white man told the Indian to come inside the station with him to the office. The Indian refused to, rightly, and after pushing him about a bit more the white man said he was going to bring someone out to him, and he disappeared inside. Of course, he didn't come out again, as I knew he wouldn't and I told the Indian so, and I also told him he was quite within his rights in all that he did and said. This doesn't sound much of an incident written about in this way, but it was rather a disgusting business to witness – disgusting that a white man should behave so arrogantly, but that is typical of the European attitude to the 'hosts', though unwilling, at whose experience we are only resident guests in Africa. We – or rather the first settlers, broke a treaty with the native Chief Lohengula when we first came. Rhodes broke it. This caused the Matabele war – and might, not right, triumphed. Do you know, Ken, that on the tea estate where I have been staying, native boys between the ages of about six and 10 do digging and hoeing in the fields for 1d per day. Others earn a tickey (the Rhodesian name for a 3d piece) and some of the more experienced boys earn 1/- a day. Apart from the absence of whips and lashes the Slave Abolition Act doesn't seem to have done much. I feel there is a great opportunity for ... a worldly ruler to come and liberate the oppressed owners of the land. Does the Atlantic charter allow for this? Does the proposed settlement of post-war England cater for these who some hypocrite once called black brothers? I have no sympathy with any political party which is solely concerned with reconstruction of England or Europe without consideration for the reformation of our own tyranny. I have written strongly on this because I feel strongly about it – probably stronger than I have ever thought about anything before. Will you come with me after the war and see Rhodesia for yourself?

I promised Cousins Apsley and Mary whom I have been staying with for part of my leave that I would come back to Rhodesia within 2 years of the cessation of hostilities to see them ... Will you come with me Ken? ... I would like you to see Africa for yourself. Will you come with me then?"

Philip died in a flying accident in November 1943 and, whilst Ken never visited Africa, his brother's views – advanced for their time – had a profound impact on Ken's perception of social justice.

At the grammar school, Ken excelled in English and obtained good results in History. Through the push for self-sufficiency in the war, and subsequently, he took a keen interest in gardening, and in both 1945 and 1946 won prizes at the Burnley Agricultural Show.

References

Spencer, Philip S-G. *The first hundred (1933 – June 1942)*. Privately published (date uncertain, but thought by Ken to be about 1947).

Spencer, Philip S-G. 1953. *African crocus and other poems*. The Fortune Press.

Spencer, K.G. 2000. *Sunny Bank School: some notes and memories*. Retrospect Vol. **18**, pp. 20-24.

Spencer, K.G. 2003. *Burnley Grammar School 1939-46: some notes and memories*. Burnley and District Historical Society.

Chapter 2
Diary for 1941
Plate 8

Ken kept a diary from 1st January to 20th August 1941. The entries bring alive the thoughts and activities of a boy of twelve, and paint a picture of a very different Burnley from that of today. Below – selected extracts:

Wednesday 1st January 1941
Stayed up 'till 4 am. last night. Amy & Gert[1] up for night. Rose at 10.00 a.m. and went into town and bought stamps[2]. Went to Padiham in afternoon with A. & G. Last night Baileys[3], Woods[4] and Nightingales came in. Five years since encounter on Coal Clough Lane with cattle.[5]

Friday 3rd January
Went to Blackburn with Phil and Dad to buy lens. Made goalkeepers

---

1. Ken's aunts – Amy was his mother's elder sister, Gert her younger sister. They lived at 18 St. Matthew Street in a terrace engraved 'Holly Mount' - a second home to Ken. Amy was headmistress at Habergham Eaves, Gert was a telephonist at the Post Office, but would have liked to be a school matron. Neither married.
2. Ken was a keen philatelist as a boy.
3. Baileys – Barbara's family. Susanna Halsted [Barbara's mother] hailed from Worsthorne; she worked as a photographer's assistant. James Bailey worked for the Parks Dept. where he designed a flower-bed in honour of the East Lancashire Regiment. He took a photograph of this for processing at the premises where Susanna worked. They married and lived at Scott Park Lodge [demolished by Mattox, Parks Superintendent, in c. 1965]; Barbara was born 3rd February 1922.
4. The Wood family lived at Woodlea, Manchester Road (the house was called that before they moved there).
5. These cattle were being driven to the slaughter-house; Ken had sought shelter in a gateway.

in evening and went to pictures with Phil. *Top Hat*. Very good. Still snowing. Car windscreen kept freezing on way to Blackburn.

Saturday 4th January

Went to Skipton. Ma and Phil did not go. Spent 6/-[6] on stamps and came back by Gargrave and Bolton-by-Bowland. Saw many pheasants and tits on the road. Amy & Gert up for tea. Cleaned out mice[7] and put stamps in book. Played football with Gert and drew 1-1. My goalkeepers are successful. Snow.

Sunday 5th January

Rose at 10.00 am. Played football for rest of morning. Rose to hear *Music While You Work*, but electricity off. Took ½ an hour to break ice on bird bath. Starkies[8] here for tea.

Tuesday 7th January

First day back at school. Morning uneventful, but went to Sadlers' Wells Opera *Hansel and Gretel* with a few of the school in the afternoon[9]. Great crisis as to who was to go as there were not enough seats. But as I paid in the morning, I got in with Gordon [Bellinger] and John [Duxbury]. Ma and Dad tried to get in but was full.

Wednesday 15th January

Got 8 out of 10 for composition and stayed in after school for Mrs Ascroft. Dirty rotter! Me and Ma on bus and conductor nearly missed my fare. Family not in evening so cleaned out mice. Barrage balloons[10] raised round Burnley - about ten I reckon.

---

[6]. 6/- = 6 shillings = 30p.

[7]. It was seeing Ken handle these mice which had first attracted Barbara to Ken.

[8]. The Spencers met the Starkies on an outing to Pendle on 15th August 1940. The two families 'hit it off'; Philip Spencer and Ivy Starkie became sweethearts and, at the time of Philip's death in 1943, were 'unofficially engaged.'

[9]. Sadlers Wells was transferred from London to Burnley as a safety precaution in the war.

[10]. Barrage balloons were floated above the town to disrupt enemy aircraft.

Friday 17th January
Cold. School. John not present. Bought more stamps. Phil returned to Colledge (sic). Old Mrs. Harker[11] asked me how I liked old Ascroft[12]. But couldn't tell her the trouth (sic) as she knows her. Went with Margaret to farm for milk and saw 15 barrage balloons.

Saturday 18th January
Rose at 10.15 and went sledging on Healey Heights with Dick. Had a few spills. Just missed a deep ditch. Rolled off sledge and it went on into ditch. Heavy snow and drifts. Hope we get snowed in as last year.

Wednesday 29th January
School. Play in afternoon *Macbeth*[13]. VG. Went in Woolworths' afterwards. Heckish thick-snow – Plate 8. Ma and Dad at *The Great Dictator*[14] in the afternoon with Hehirs[15]. When she and Dad returned home the snow came up to her waist and great drifts are forming. Cleaned-out mice.

Thursday 6th February
Snow so deep that I could not get to school this morning and I played football and cleaned steps for 2^d and made snow fort. Went in afternoon and did not take gym. Went to meet Ma at Gert's after school and made A1 fort in their garden, complete with moat and tunnels and paths. Thaw has now set in and is quickly clearing snow.

---

[11] The Harkers were related to Ken's father by marriage – his sister had married into the Harkers who ran a hardware shop.

[12] Mrs. Ascroft taught French. Her husband was a founder member of Burnley Civic Trust.

[13] The Old Vic theatre – transferred from London to Burnley for the duration of the war.

[14] Charlie Chaplin skit on Hitler.

[15] Mr JAB Hehir – Headmaster of Rosehill School, lived in Coal Clough Lane.

Monday 10th February
Lucky Day. School. In afternoon met my girl friend AND SPOKE TO HER.

Thursday 13th February
Play Twelfth Night in afternoon with school. VG. Gave 2^d to hospital collection. Saw girl-friend on bus down to school in afternoon and made plan to take her photograph.

Saturday 15th February
Won 2/6[16] for joke in 'Express'. Shall split it up among us 4, as Brian has joined our gang.

> *"Mister"* said an urchin to the man, who was driving a very poor and thin horse, *"do you want me to hold him?"* *"Naw, this horse t'll not run away."* *"Aw doesn't meean to hold him fast sooas he'll not run away. Aw meean hold him up sooas he'll not drop."*

Wednesday 26th February
Phil was born on this day in 1922[17]. Ma and Dad at early service at St. Matthew's to commemorate the event. Cleaned-out mice. Tried at all shops in town for Balkan Zobranie cigs for Phil's birthday, but either heavily taxed or sold out. Bees' seed book arrived.

Thursday 27th February
School. After school met Dick and bought Phil's present, a tie; and seeds and an identity disc for Margaret.

Friday 14th March
Got a dashed lot of homework for Monday. Scripture, History, Geography, French, English and Physics, Algebra and Latin. Not forgetting Chemistry. 9 altogether <u>and</u> have to write to Phil and Brian. Alert at 9 pm. I stayed up till 10.30. Slept in front room. Planes over and bombs somewhere.

---

[16]. 2/6 = the two-shillings-and-sixpence coin was known as half-a-crown and would be worth 12.5p today.

[17]. 'This day' means Ash Wednesday – Philip's actual d.o.b. 1st March

Sunday 16th March

New hen laid 2 eggs, 1 in morning and 1 at 4 pm. Gert and I went walk round fields hoping to see George[18], but being such a good fighter he was locked up.

Friday 21st March

School. I weighed 5 st, 4 lbs, 8 ozs + was 4ft 10¼ in high. Increase of 2½ lbs since last time.

Tuesday 25th March

School. Games in aft. Ran like heck after bad start & fell in mud & came in 6th. Thought it was cross country trials & I had done course in time but was only a trial. Swoon!

Monday 31st March

School for afternoon. In afternoon Ma gave me 2/- of which (she said) I was to spend 1/6. Well I spent the lot so the old girl raised the house. She took my stamps off me and said I could have them back when I had returned her the 6^d.

Saturday 12th April

Rose at 9 am. and did all errands for Ma in Thomas's[19] & Rotherham's[20]. A & G at Worsthorne with Dad so took goods home in relays and nearly dropped them.

Tuesday 15th April

Rose early and went cross country over Hambledon. Got back at 7.45 and beat Phil easily.

Wednesday 16th April

Helped with perfect ciné show at Habergham School to expand guild. Phil said I helped a lot.

---

[18]. A litttle black bantam.

[19]. A baker's shop in St. Matthew Street.

[20]. Park Dairy – a newsagent-cum-sweets-cum-cigarettes. Mr. Rotherham once opened packs of cigarettes for Ken just to find a 'rare' cigarette card for him.

Thursday 24th April
Got up at 10 am. & Phil went back to college. Went to fetch Barry to stay with me. Took me 15 mins to get there [on bike] and 90 mins back! [uphill]. Went to bed at 10 pm and talked in bed.

Saturday 26th April
Rose at 6.30 am. and went to Margaret's room dressed in towels and masks. Gave her a good shock. Before breakfast caught tadpoles at Crown Point. Went to Nightingales for dinner and got more tads and snails at a pond round there.

Sunday 27th April
Rose at 10.30 am and was going to go for walk with Gert but did not. Gert's meat bad – a horrible smell[21], so had lunch at home. Left tads & snails at Gert's. Cleaned mice and slept at Gert's & heard Winston Churchill.

Monday 28th April
Bike broke down on way to Skipton, but met Tattersall, who mended it at Colne. We went to Earby. Got stamps at Skipton and Gargrave. 2d blue for 1d also more at Nelson.

Wednesday 30th April
School. Got tons of homework. French test dammed swizz. Swell lunch; chocolate at tuck shop. Marvellous. Just got done out of a stamp at art shop by Mullen. Saw Tattersall and went to Woolworths and saw good stamps. Amy here and set all my seeds & potatoes. Must get more land.

Saturday 3rd May
In afternoon run to Barley near witch's cottage and raced twigs in river near stepping-stones with Gert & Ma. Ma won, me second, and Gert last. Alert & guns. Bad raid Manchester.

Tuesday 6th May
School. Went on bike to running in afternoon. 3rd place. 8 bombs dropped on Rossendale Road last night. Many windows broken but

---
21. No home refrigerators in those days.

Joan's [Riley] house untouched. In evening bombs woke me & slept in Ma's bed.

Friday 9th May
School. Tracked Joan Riley[22] to house, which is not the one I thought, but near, if not Burwains. Williams[23] away, so no gym though it didn't matter to me. No alert for first time for a week.

Saturday 10th May
Collinges' cows out for first time.

Monday 12th May
Went on bike and found name of house to be Moorcroft. Joe Key[24] (T.R. Key) left to join the RAF.

Tuesday 13th May
School. Won at running at last, a tie for 1st place though had to wade across river to get it.

Wednesday 14th May
Coming up on bus in afternoon saw Gert & JR. JR was showing books to friend and saw her name and form same as me. 3B. She's jolly good at English, though could not spell 'occasionally'.

Monday 19th May
School. A good & exciting day. Bought a rhubarb root with the 2/- Dad gave me last night. Got manure in eve and went past JR's house, luckily before we got manure and she came out and crossed the road and went up Ross Avenue. Wow, what a thrill. Dammed exciting.

Tuesday 3rd June
Quite a good day. At Gert's all morning & pictures & *Tin Pan Alley* with

---

[22] Joan Riley – a schoolgirl Ken fancied, but never even spoke to.
[23] Williams, P.E. instructor, went from B.G.S. to become an R.A.F. P.E. instructor in Africa.
[24] A teacher.

Gert at the Palace all afternoon. Walk and saw darling Joan in evening on a path ... gee but she's swell.

Wednesday 4th June
Saw Joan on her bike in evening & hurt my leg rather badly. Later rang up Ma at Mechanics and she came up and helped me. A nasty accident on bike. JR saw me too, I think.

Thursday 5th June
Not much to say except went to [Dr] Shotton[25] to have bone set for it was twisted.

Tuesday 10th June
School Sports in afternoon ... Ma & Dad & Amy there in car. In Habergham Clough with John in evening. Saw cuckoo very closely and rode all way up Rossendale Road till home after flying paper gliders. Got sketch of sunset over Nick of Pendle at 9.30 pm.

Wednesday 11th June
School. Walk with Margaret in evening. Saw and got cheek from Rose ... she was dammed bold. Also saw Geoffrey (Thornber) and our old mines (by farm – iron pyrites). On bike at school in afternoon and got caught on road by cows one of which jumped a bridge. Joey[26] caught a young owl at school. Baby birds in our box (house sparrows).

Saturday 14th June
Gee, but what luck! (Good too) Got a picture of Joan from Burnley Express. She took part in some play for W.V.S.

Sunday 22nd June
In long pants for first time.

---

[25] Dr Shotton was an osteopractitioner on Barden Lane. In getting off his bike, Ken had twisted a ligament in his knee.

[26] School caretaker – tawny owls nested in ventilator at Grammar School every year.

Tuesday 24th June
School. Birthday & got 10/- (= 50p) as well as many presents. John, George & Gordon here for tea and had presents & cake. Came to Aunties and watered garden. All plants doing well.

Friday 27th June
School. Cold just put off then Ma Ascroft opened all windows and perfect drafts (sic) caused cold to come on again.

Monday 30th June
Watched haymakers and went to bed early and read a bit.

Wednesday 2nd July
Was lucky and got a cos lettuce for 5^d this morning. After dinner (of lettuce) Phil and I rested in fields and saw a Lysander[27]. With Gert at 4 pm. to see cows come in. Saw the same Lysander come low and circle at 9.00 pm. Coincidence!

Wednesday 9th July
Did garden and then went run to Hurstwood with Baileys. Got shoved mostly by Phil & Barbara. Beastly. Got some good ghost stories[28] off Mrs. Bailey however.

Wednesday 30th July
At Empire in the eve with Ma, Dad & Phil. Then got caterpillars [poplar greys] at Drill Hall.

Friday 1st August
At Hurstwood with Phil & A in car in afternoon. Was chucking stones at empty tins in the lake[29] with Phil. Found 4 puss-moth caterpillars on way back.

---

[27]. The Lysander - Second World War aircraft.
[28]. Spine-chillers about the Bay Horse Hotel
[29]. Not the reservoir but a small lake on way to it.

Wednesday 6th August

Went with Starkies to Whitewell. Found an oak eggar on heather at Whitewell. Chased sheep with Gert. Had dinner out. Barrages visible from Whitewell. Tea at Parker Arms Newton.

Wednesday 20th August

Rose at 9.45 am. and read Picture Post. At Phil's St. Matt's show + 100 kids there and raised 10/6. Heard owl in park then Ma came in and I retired at 10.30 pm.

Chapter 3
Down the Pit
Plate 10

In December 1946, Ken volunteered to go down one of the coal mines near Burnley as an alternative to National Service. This enabled him to continue with his bird studies. He was what was known as a 'coal mining optant', i.e. he had chosen to do this as distinct from a 'Bevin boy' who had been conscripted into the pits.

He signed on at Bank Hall Colliery (Nadin, 2003) where he worked until he was demobilised. He found the miners to be a congenial group who, at that time, still had their own dialect, some of which Ken noted (Spencer 1980). One or two words remain in general use, such as 'bait' for packed-lunch – typically jam butties - carried in a 'bait-box' to protect them from the rats and mice which lived down the pit.

A collier friend, John Nuttall, used to recount his observation that a large proportion of these rodents appeared to be born blind. Just how far these animals had evolved in response to the darkness of the pits will never be known since, following its closure, Bank Hall has been filled-in.

Ken can be seen as he was in 1947, aged 18, outside the back door of his Aunt Amy's & Gert's at 18 St. Matthew Street, wearing his 'pit clothes' which came from the local Army & Navy Stores (Plate 10).

Terms of interest recorded by Ken include:

- "check – a numbered disc handed to each worker as he went underground. Those first-in got a low-numbered disc and so on. At the end of the shift, those with low numbers were given priority when coming out;
- tally – a disc to identify a full tub as 'belonging' to the collier at the face with whose coal it was loaded;

- tram – a flat wooden board, like a sledge on wheels, used by workers who had a long way to travel from the shaft bottom."

References

Nadin, J. 2003. *Collieries of north-east Lancashire.* Stroud.

Spencer, K.G. 1980. *Some memories of working down the pit, Burnley 1947.* Journal of the Lancashire Dialect Society, **29**, 23-25.

Chapter 4
Ornithologist
Plates 11 - 13

A snippet of paper dated March 1933, which Ken found amongst his Mother's treasured possessions, is the earliest record of his attraction to birds. In bold block capitals he had written 'I LIKE BIRDS'; he would have been five years old.

His observational skills were built on this early interest. A letter to the Burnley Express dated 5th February 1944 reads: *"I would like to report the first song of the skylark in Burnley district this year. I observed a pair in full song on Hambledon on January 30th, which is the earliest date I have yet recorded."*[1]

Influential in Ken's life at that time was Clifford Oakes, a Burnley man and author of *The Birds of Lancashire* (1953). Clifford ran a pie shop close to Turf Moor, Burnley's football ground. On his days off, Clifford cycled around the county, noting his observations. In modern times, Ken considered Clifford could have aspired to be a professional ornithologist. Ken recalled: *"I shall always remember how, in 1944, he sought me out as a schoolboy and, much to my delight and pride, asked if I would like to help him with the Rook Survey ... He taught me to be a ringer and introduced me to the British Trust for Ornithology, the Royal Society for the Protection of Birds and the British Ornithologists' Union."* Ken joined the RSPB in 1945.

Between January 1948 and June 1952, Ken estimated that, in the course of his researches on the lapwing *Vanellus vanellus*, he spent "thousands

---

[1] Contrast this with recent times. In a letter to Susan Halstead in May 2009, which included a copy of his bird records for 2nd April 2009, Ken wrote *"The Silent Spring. No lapwings on the Tewit Fields round Crown Point for the first time in my experience of 65 years. Very few skylarks or meadow pipits. Snipe and twite, partridge and redshank now completely gone."*

of hours in the field and ... walked fully 3,400 miles."' During this time he frequented *" ... a fair stretch of high meadowland, plough, and bleak windswept pasture ... except for the winter observations, which were made on the low green fields to the south-east of the town."* He spoke about the breeding behaviour of lapwings at the Oxford Conference on Bird Biology 1950.

The eventual result of his fieldwork was *The Lapwing in Britain* (1953), written whilst at Leeds University, which received excellent reviews (see box, also flyer – Plate 12). Tate (1979, p.109, & 202-203) refers to it as a classic account of the lapwings' social history.

Whilst *The Lapwing in Britain* was, and remains, Ken's single most significant publication, it was by no means either his first or his last. He began submitting observations to *British Birds* as a schoolboy in 1946. *"Several miles from Burnley there is a very large roost of starlings* Sturna vulgaris *situated in a conifer plantation. Here, in the evening, as the starling flocks pour in from all directions, one may see a small number of kestrels soaring and gliding above the incoming birds. Very shortly, one of the former will fly down into a large gathering of the smaller birds, scattering a few from the outer edge of the 'cloud' and pursuing one selected bird. The chase may continue over a good distance, and often the fugitive escapes by rejoining the main flock. Occasionally, however, the kestrel will, either by luck or by superior skill, capture the starling in mid-air and glide down into the trees to consume its prey. I have seen as many as 3 birds brought down in this way during the course of a single evening."* With his submission (2009), concerning blackbirds *Turdus merula* eating *Fuchsia* seeds, Ken has been submitting observations to *British Birds* for more than 60 years.

In 1951, Ken submitted an account of a pair of rooks *Corvus frugilegus* which built an isolated nest which was subsequently destroyed by other rooks at a time when it contained at least one squab, the latter being found dead beneath the tree. *"I had often heard the story that when rooks build far apart from their neighbours the latter destroy the nest, but*

up to 1950 had thought it just another 'yarn'. Now I'm not so sure ..."
When Cocker (2007) came to write his masterpiece Crow Country, he thanked Ken as one of the great polymaths of the bird world for *"chasing down ... such a wide sweep of rook-related materials"*.

I have vivid memories of Ken drawing my attention to the thin 'seeip' of redwings *Turdus musicus* passing over Burnley on still autumn nights. In a note published in 1952, he gives a detailed account of the migration of these Scandinavian thrushes during the night of 18th - 19th October 1947, when he counted individual calls with a view to determining the intensity of their passage. The meticulous nature of his observations may be judged from the following extract:

> *"Met conditions ... fair, slightly foggy, overcast ... sunset 17.04 sunrise 06.28 GMT.*
>
> *I made three series of observations [i] 19.00-23.00 [ii] 03.00-03.30 [iii] 06.00-07.00 GMT.*
>
> *During the first watch, passage was intense, calls per ½ hour numbering 25, 41, 35, 65, 46, 31, 35, 42.*
>
> *By contrast, between 03.00 and 03.30 only 1 call was heard and it was clear the passage had virtually ceased.*
>
> *Counts at 06.00-06.30 showed a revival, 32 calls being heard but, between 06.30 and 07.00, movement was evidently slackening again, for I only noted 8 calls. The birds themselves were visible this last half hour - 4 flocks ... with about 20 birds in each, The fact that so few calls were heard from these suggests that the number of redwings which cry out whilst within the listener's earshot may bear small relation to the numbers actually passing. I have since confirmed this suspicion by some observations on visible flocks at dusk, finding that only one or two 'seeips' are usually heard from each lot as they go by ..."*

My favourite bird – the Lapwing

by Ken Spencer

from Young Naturalists' News 1981. British Naturalists' Association.

My auntie Amy was a retired teacher. In her home she had two or three children's reading books and I was very fond of one of them in particular. It had some simple stories in it about birds, illustrated with coloured pictures by a first-class artist, A. W. Seaby. One of them showed a mother lapwing brooding her little family of chicks. The story that went with the picture told about a young boy walking out over the fields and coming near to a lapwing's nest. Naturally this worried the mother bird, but by pretending to have a broken wing she very cleverly decoyed the lad away from her eggs.

I liked that story and was reminded of it when, much later in life, I did really come upon a lapwing which played that trick. I found her tiny chick crouching in the grass on a field near Stromness in the Orkney Islands. I knelt down to look at it closer and the mother came right up to within three or four feet, calling excitedly and flapping about as if crippled. I had our old family camera with me and tried some snapshots. They didn't come out very well, but at the time they were the only pictures ever taken of a lapwing performing distraction display and I was quite proud of them.

Let me tell you how I came to study lapwings specially.

When I was about seventeen, I remember having one particularly bad cold, which kept me in bed for days. While I was there, I read David Lack's superb little book *The Life of the Robin* and I could hardly wait to get up and about again bird-watching. As soon as I could, I was out watching robins – but I found that there were so many of them in the local park that I got confused as to which was which. I wasn't keen on colour-ringing, which might have helped me with that, so I switched instead to an easier species, the song-thrush. Unfortunately, when the nesting season came round, all the nests I knew of were destroyed by young vandals and so that aspect of my study came to an end.

Luckily, a friend asked me to join him that year for a holiday in Orkney and it was there that I really got started on the lapwings. Using a borrowed bicycle, I spent whole days exploring the country roads and watching lapwings, curlews, redshanks, harriers, short-eared owls and so forth. I can remember staying up late at night and getting up very early because I didn't want to miss a minute of the day.

When I came back home, I found I could continue my lapwing studies. The birds were fairly common locally and they were not too badly harassed by the bird-nesters, so that over the next four or five years I was able to spend literally hundreds of hours watching them. I used to be up on the fields at all hours. Many people only go bird-watching in the afternoon, but I always found the first two hours after sunrise and the last hour before dark were the best.

People would laugh if you said you were going out to study birds at night, but in fact some birds are very active then. Owls are an obvious example, but lapwings too can be very lively after dark. When the moon is full, they feed at night and converge into flocks for sleeping during the day – which is why the biggest lapwing flocks are usually seen at full-moon times.

The lapwing is a lovely bird. It is lovely in itself, and lovely in the places where it lives. You will understand why I call it my favourite bird.

THE LAPWING IN BRITAIN
Plate 12

The author of this book is at present employed on the staff of Leeds Public Libraries. Recently described as "a young ornithologist of considerable promise", he has spent literally thousands of hours in the hills adjacent to the county border near his hometown of Burnley, Lancashire, making the field observations here recorded for the first time.

EXTRACTS FROM A FEW PRESS NOTICES

- "Full of painstaking work and minutely documented. A good example of what can be done by a disciplined and assiduous worker in his own neighbourhood." *The Times Literary Supplement.*

- "I can thoroughly recommend *The Lapwing in Britain*; Mr. Spencer's enthusiasm is fine and his interesting book a real contribution to our understanding of natural history in general and a lovely wader in particular." *James Fisher, Country Fair.*

- "An important monograph ... a book which is not only a valuable contribution to ornithological science, but which also makes entertaining, stimulating and extremely profitable reading." *Philip E. Brown, Ibis (Journal of the British Ornithologists' Union).*

- "In this present work, illustrated by some beautiful photographs, Mr. Spencer has brought together a vast array of facts and conclusions about this particularly attractive bird, whilst such matters as voice, flocks, roosting and migration, instead of being dismissed in a few words, have whole chapters to themselves." *Countryside.*

- "A valuable addition to the literature of British ornithology." *W.B. Alexander, The Naturalist.*

- "Even those who have no interest in the biology of birds can hardly fail to be captivated by the chapter on the Lapwing in British folk-lore." *J.K. Adams, Country Life.*

- "Practically everything that is known about the Lapwing has been drawn together in this volume." *Bird Study (The Journal of the British Trust for Ornithology).*

- "At once a serious work for ornithologists while holding great interest for the general reader." *Yorkshire Evening Post.*
- "A detailed study of one of our most popular, beautiful and useful birds." *The Countryman.*
- "En somme une compilation très approfondie à laquelle l'expérience personelle de l'auteur donne un accent original." C. Jouanin. *L'Oiseau et la Revue Française d'Ornithologie. Paris.*
- "It is a solid work which will be useful for reference, and with its exhaustive citation it also combines much first-hand observation." *British Book News. (Published by the British Council for the National Book League.)*
- "We much look forward to further work from Mr. Spencer, who is an ornithologist with a future." *Glasgow Herald.*

In 1975 Ken published his observations of breeding house-martins *Delichon urbica*, but it was not so much their behaviour at the nest which interested him as the question of where they roost when not sleeping in the nest. This he called 'the great question'. He became convinced *"... that most house martins, when on migration, sleep in trees, in a way which could be described as loosely communal"*; trying to prove this beyond doubt in poor light eluded him. His brief paper (1978) ends with a challenge to solve this mystery.

As with the lapwing, Ken was attracted to the Pennine population of curlews *Numenius arquata*. "Every year the home-coming of our curlews from their winter quarters on distant marsh and mudflat is an eagerly awaited event. At any time from about February 18 to March 20 it may be expected ... curlew display still remains something of a mystery ... departure from the hills begins just before the end of June ... WSW towards the Ribble estuary is the direction that most birds take ..."

Since 4th January 1945 Ken has kept a daily diary of birds seen. From 1957 to 1978 he was editor for the *Lancashire and Cheshire Fauna Committee* of the annual *Lancashire Bird Report*. This provided him with the basis on which to note the changing status of birds in the county.

In 1998, Ken heard four chiffchaffs *Phylloscopus collybita* singing in Towneley Park, whereas Clifford Oakes (1953) had heard only one in his whole lifetime! In 1999, long-tailed tits *Aegithalos caudatus* nested either in or very near to Scott Park – *"this would have been like a dream to Clifford or indeed even to me in my early days, when they seldom occurred outside the Ribble Valley."* In the same year ravens *Corvus corax* nested successfully in the Cliviger valley for the second year in succession. However, *"we are losing redshanks ... and curlews as well."* New arrivals on the Leeds and Liverpool Canal in the vicinity of the Weavers' Triangle included goosander *Mergus merganer* (December 1999) and coot *Fulica atra* (March 2009).

A Pennine speciality used to be the twite *Carduelis flavirostris* but, by September 2002, whilst they had virtually gone from north Rossendale, a population to the north-east of Burnley was still doing well. United Utilities engaged Peter Grice to make a special study of them. Around Rossendale, Crown Point and Clowbridge, twite had disappeared, but at Hurstwood above Worsthorne they were hanging on. One of the duties undertaken by United Utilities' warden, Linda Williams, was to put seed out for the twite throughout the winter (*Lancashire Evening Telegraph*, 13th December 2004). She recounts: *"One day I saw two twitchers laughing at me and when I asked them why they said about 80 twite had followed my car to the feeding area".*

During the late 1950s the *East Lancashire Ornithologists' Club* (ELOC) organised a different survey each year. That for 1957 focused on swift nest-sites. Writing in 1959 Ken said *" ... one of the very few authors accurately to describe the habitat of the swift* Apus apus *as it is hereabouts, and must indeed be throughout a great part of industrial Britain, is Clifford*

Oakes (loc cit), who points out that although there are of course some rural nesters, nevertheless 'As a breeder the swift is thoroughly urban ... for it nests abundantly in the heart of manufacturing towns, and indeed the majority of the breeding population are not merely town-dwellers but town gamins, chasing and screaming around their nesting sites in mill chimneys and dingy warehouses miles from open country.'"

In 1959 Ken submitted that "the breeding population of the swift in the industrial north is now as great as at any time in the past." In what has proved to be prescient he went on "... there are signs that it may have reached its limits ... Food supply ... appears to be the crucial factor and it seems with improved sanitation and extended use of insecticides – not only in this country, but in those where the birds winter – the situation from the swifts' point of view is unlikely to improve." The use of insecticides on crops may be a primary reason why the accumulation of insect 'gunk'[2] on car windscreens is now a thing of the past. This may be indicative of a significant reduction in aerial plankton available to swifts and hirundines.

At the *British Ornithologists' Union Conference* in Bath in 1962, Ken presented a paper 'The roosting behaviour of starlings.' His work had been inspired by Arthur Welch, himself something of a specialist on starlings *Sturnus vulgaris*. Arthur's pioneer work on the communal roosting behaviour of that species at Barley-in-Pendle from 1943-47 paved the way for Ken's own studies. Ken subsequently published (1966) an account of the roosting behaviour of starlings. When Moss (2008) wrote about starling roosts, Ken was pleased to have been cited.

Always a careful observer, Ken noted an increase in mallards *Anas platyrhyncos* at both Towneley Pond and on the Leeds-Liverpool Canal. He first recorded them on the canal in 1980 and at Towneley in 1984 but, by

---

[2]. The smashed bodies of insects which used to smear car windscreens in summer to such an extent, and which were so resistant to windscreen washers, that one sometimes had to stop and clean the windscreen en route.

1999, numbers had built up to 110 and 130 at these respective locations. Gradually, with this increase has come a numerical preponderance of males to females with a ratio of 3:2. Ken speculated that this may be a form of population control, with 'gang rapes' occurring which can be fatal to females. He further observed in May 2006 that *"no species has females which differ so much in character individually. Some are what we would call good mothers, others couldn't care less. There are tiny ducklings on the canal right now which are practically bringing themselves up."*

Writing of house sparrows *Passer domesticus* (in Caul 1968) Ken described them as *"so numerous as to need no further mention"* but, by 2000, they had declined such that he was *"keeping a note of almost every sighting, something that would have been both ridiculous and almost impossible in past years."* In the 1960s a population of c.40 frequented the gardens around 167 Manchester Road but, by 2008, these had dwindled to eight, and by 2019 to nil.

In a note for 1945, Ken said of Crown Point *"cuckoos everywhere"*, whereas in 2006 he noted that *"they have completely gone."* The one factor which, it seemed to Ken, had changed was what he termed 'photochemical smog'. Following just one fine, still day over the town, this polluting layer, which he believed to be largely derived from vehicle exhausts, could be seen as a purple haze. This, he felt, must be toxic to the invertebrate-life so important to many fledgling birds.

Ken maintained a life-long interest in lapwings. His letters reveal a progressive diminution of the local breeding population. Disaster struck when lapwing fields at Crown Point were spread with slurry at the height of the breeding season in April 2002. In one of his more recent observations in May 2008, Ken thought carrion crows taking eggs and harassing adults might be partly to blame for their decline.

In May 2002, Ken described the decline of local birds as frightening. *"I'm thinking of curlews, lapwings, redshanks, yellow wagtails, yellowhammers.*

Where the fields are what you call nitrous green[3] it is understandable, but even when I find an old-style field the birds aren't there. And yet some species, mainly woodland ones, do well: willow-warblers and chaffinches are numerous: goldfinches, too."

A close friend of Ken's, Eric Ward, took part in a study of the declines of ground-nesting birds including lapwing *Vanellus vanellus*, snipe *Gallinago gallinago*, skylark *Alauda arvensis*, twite *Carduelis flavirostris* and reed bunting *Emberiza schoeniclus* at two sites in the south Pennines (Fuller et al, 2002). Progressive changes in land-use, involving loss of rough grassland and a switch from dairy to sheep-farming, may have contributed to the declines at one of the sites. However, there was no obvious change in land-use or habitat at the other site where population declines began 5 to 10 years earlier. Such declines have probably occurred widely in moorland-edge areas during the last 30 years and multiple factors may have been responsible. By June 2006, the outlook for lapwings was bleak and for the second year in succession he could see no evidence of breeding success at Crown Point.

Ken took encouragement from work being undertaken by Anne Heslop for the RSPB's Bowland Lapwing Recovery Project. To give her an idea of its former status, Ken sent her a copy of Clifford Oakes' field notes from the *Burnley Express* of 8th January 1938: *"I was able to spend a considerable amount of time on a special investigation into the habitat of the lapwing, organised by the BTO. My share in this task was a road transect comprising Burnley, Whalley, Mytton, Higher Hodder, Red Pump and Bashall Eaves, a distance of approx. 17 miles. I covered this route twice a week by bicycle throughout the season, watching 40 pairs of lapwings from the egg stage to the time when the young joined the adult flocks. Fully half of these nested on rough lowland pasture, the remainder in varying proportions on meadow, rushy fields, ploughed land, and gorse-*

---

[3]. My term for heavily fertilised pastures which turn an intense bluey-green.

covered common. I afterwards computed the status of scattered flocks in parts of the district aggregating 1,400 birds."

The impact of modern grassland management on lapwings has been devastating and it is only when a farmer is sensitive to their needs, such as Robert Greensit, who was declared *Lapwing Champion for the North of England 2007* (Johnson, 2007), that populations locally approach levels which were commonplace back in the 1950s.

In the ELOC report for 1997, Ken described the habit of lapwings frequenting the extensive gently-inclined, pale, roofs of some of the modern warehouses which are now so extensive around most towns and cities. Calbrade *et al*, (2001) considered the use of the large roof areas of commercial and industrial buildings for daytime roosting by lapwings *Vanellus vanellus* and golden plovers *Pluvialis apricaria* as *"a relatively new phenomenon, confined mostly to northern England."* Ken described seeing a flock of lapwing disperse to their feeding grounds at twilight as a *"curious experience ... standing by a roundabout with scores of cars going past ... and seeing a very 'wild' sort of event."*

In August 2015, Ken was surprised and delighted to receive a 'Gold' BTO lapel pin from Andy Clements, BTO Director, by way of thanks for his loyal support through his BTO membership dating from 7th January 1947. See Annex.

Some bird events lost in recent years: a note by Ken, February 2017

- House sparrows: flocks in fields and in roosts.
- Starlings: daily flight lines to communal roosts.
- Curlews: returning to the shores after nesting in the hills; flying high overhead in the early evenings, their presence betrayed only by their calls from high above.
- Twites: flights of twos and threes to waste ground in Burnley from the nesting fields.

- Willow warblers: passing via Burnley gardens after nesting.
- Redwings: groups – almost always of c.10-25, moving west over St. Matthews in early daylight and their thin 'seeip' calls to be heard as they pass overhead during hours of darkness on still autumnal nights.

THE LAPWING

K.G. Spencer, RSPB Birds Magazine 1972, pp. 121-124.

In late February or early March there often comes a day of outstanding beauty, such as Emily Brontë described in *Wuthering Heights* *"The snow is all gone ... the sky is blue and the larks are singing and the brooks are all brim-full."* I always look forward to this day for it is likely to bring the lapwings back to their Pennine fields and, even in the towns, we have only to look up to see flocks travelling over, strung out in lines across the cloudless sky.

If I go up to the hills that evening, I expect before long to find a group standing quietly in the fading light. I take up my position at the corner of the stone walls and, at about forty minutes after sunset, I watch the males fly out and go to their separate territories, where they perform their aerial displays and sing: *"Ayhrre willuch-o-weep, weep, weep; eyuweep"*.

Song-flight, the males' most obvious form of territory advertisement, is a feature of every lapwing field in early spring. So, too, are the aerial clashes which repeatedly occur where boundaries adjoin. It is interesting to note that territories are almost certainly three-dimensional. That is to say, they are defended lengthwise, across, and also upwards to a certain height. Above that height, a passing lapwing may travel by unchallenged.

Displays diminish when the eggs are laid. The females do most of the incubation, with the males on lookout duty nearby; when relieved for an interval the females usually go to feed at a neutral point outside the territories. Four eggs is the usual clutch. On the evidence that I was able to gather pre-1952 it seemed likely that smaller clutches were commoner in the north, which is the reverse of the usual rule for birds. I am waiting with interest for someone to analyse the *British Trust for Ornithology* nest record cards, which are now sufficiently numerous to give a statistically valid result.[4]

When the chicks appear, both parents unite full-time in their care. Though only the female broods them, the male is always close at hand on guard. A skilled observer can assess the approximate stage of a brood's development by the intensity of a parent's reactions to him. When with full-grown chicks, both may be almost apathetic. Indeed, if it is getting rather late in the year, one parent may entirely desert the family and join the nearest post-breeding flock.

Such flocks begin to gather earlier than is generally realized; late May is not exceptional. In Norfolk, lapwings are actually on the move by that time, en route for the European mainland. Richard Perry has described the scene clearly: *"At the end of May small flights are already flickering low across the marshes day after day: often skimming the ground in leisurely flight, but always making steadily west. A dozen such small flights go over in a day, some of only eight or nine birds, others of seventeen or twenty-five ..."*

On the Pennines, our flocks remain on high ground until well into autumn, and the shallow-sloping embankments of bleak reservoirs are favourite haunts where hundreds assemble to bathe and preen during their moult. Lowland flocks reach their maximum numbers in October and November, after which hard weather usually drives most of the population to west Lancashire, Ireland or the continent.

Flock routine is governed by the moon. When the moon is full or nearly so, the birds rest by day and feed by night. At such times, flocks which

---

[4]. See *The Lapwing* by Michael Shrubb. 2007. T. and A.D. Poyser, London.

for most of the month are widely dispersed about the countryside will congregate at a favoured day-roost. When I wrote in 1952, the largest flock reliably recorded in Britain totalled 15,000, counted by Dr. Bruce Campbell and the late A.W. Boyd at Witton Flashes, Cheshire on 25th November 1950, the day after a full moon. Mr. M. Jones has since recorded a flock of similar size near Cockerham, Lancashire, on 6th February 1966, again on the day after a full moon.

A single-species study is extremely time-demanding. As the years have gone by, other interests and commitments have come into my life, and lapwings have unavoidably been crowded out. And yet, as I sit here now, on a sleety night in midwinter, I find myself thinking forward once again to that day in February or early March ...

A lifetime's summary of the main changes amongst East Lancashire birds: a note by Ken, February 2017

Of course, there have always been changes and always will be. Around Blackburn for instance, Skaife (in 1838) described the whinchat as *"common in every meadow"*. We can hardly imagine that today. There was even a public house called "The Utick's Nest". Everybody knew what an Utick was.

I was born in 1928. Then, and for many years afterwards, lapwings, curlews and redshanks were familiar nesters. The *Lancashire Wildlife Trust* took the lapwing as its logo.

As we all know, these have declined to the point of regional rarity. They do well only on nature reserves or on keepered land, which in itself points to what I am sure is the main cause of the loss: the increase in carrion crows and foxes. Carrion crows first nested near Burnley in 1952; now they are everywhere.

Again, through most of my lifetime, insectivorous birds like spotted flycatcher and yellow wagtail were always here in summertime. Now they have almost gone. There has been a massive decline in airborne insects. It was once common for windscreens to be splattered with their remains. No longer so. I put this down mainly to the serious advent of vehicular smog c.15 years ago. When there is rain, the particles that make up this smog must be depressed onto the earth, with who knows what effect upon invertebrates? I am surprised that so little research has yet been done on this.

What I have written is only an outline sketch. I haven't touched upon the changed status of some other birds – cuckoos, twites, skylarks, Canada geese, nuthatches, goldfinches, jays, jackdaws. Some you win, some you lose.

Thanks to Dr. S.D. Ward, I have kept up fairly well with the current literature and I know that other factors in declines are involved that I haven't mentioned. I merely felt that in my limited way I should draw attention to what may be factors that so far have been under-recognised at top level.

In 1969, at the University of Salford, Ken was awarded the degree of MSc in Environmental Science, for which he wrote a dissertation entitled 'East Lancashire reservoirs: an ornithological survey.' The degree was presented by the Duke of Edinburgh.

References

Calbrade, N., Entwistle, C.A., Smith, A.J. and Spencer, K.G. 2001. *Roof assemblies of lapwings and golden plovers in Britain*. British Birds. **94**, pp. 35-38.

Caul, R. (Editor). 1968. *The natural history of the Burnley area. Birds* by K.G. Spencer, Burnley Corporation.

Cocker, M. and Mabey, R. 2005. *Birds Britannica.* Chatto and Windus, London.

Cocker, M. 2007. *Crow Country*. Jonathan Cape.

Fuller, R.J., Ward, E., Hird, D. and Brown, A.F. 2002. *Declines of ground-nesting birds in two areas of upland farmland in the south Pennines of England*. Bird Study **49**, pp. 146-152.

Johnson, H. 2007. *Where farming and wildlife conservation work hand in hand: Helen Johnson meets a farmer who has been declared this year's 'Lapwing Champion for the North of England'*. Dalesman November, pp. 54-58.

Lack, D. 1962. *Migration across the southern North Sea studied by radar. Part 3. Movements in June and July*. Ibis, **104**, pp. 74-85.

Moss, S. 2008. *A sky full of starlings: a diary of the birding year*. Aurum, London.

Oakes, C. 1953. *The birds of Lancashire*. Oliver and Boyd. Edinburgh and London.

Perry, R. 1938. *At the turn of the tide*.

Spencer, K.G. 1946. *Kestrels capturing birds in full flight*. British Birds. **39**, pp. 216-217.

Spencer, K.G. 1947. *Starling roost*. Countryside. **14**, pp. 40-142.

Spencer, K.G. 1949. *Behaviour of young and adult lapwings*. British Birds. **42**, pp. 26-27.

Spencer, K.G. 1949. *Observations on robins in late winter*. British Birds. **42**, pp. 184-5.

Spencer, K.G. 1949. *Autumn aerial activity of kestrels*. British Birds. **42**, pp. 390-391.

Spencer, K.G. 1951. *Observations on a pair of rooks in an isolated nest*. Bird Notes, **24**, pp. 160-162.

Spencer, K.G. 1952. *Observations on a pair of tawny owls.* Bird Notes, **25**, pp. 141-145.

Spencer, K.G. 1952. *Nocturnal movements of redwings.* British Birds, **45**, pp. 367-368.

Spencer, K.G. 1953. *The lapwing in Britain: some account of its distribution and behaviour, and of its role in dialect, folk-lore, and literature.* A. Brown and Sons Ltd, London & Hull.

Spencer, K.G. 1959. *A survey of swift nesting sites.* The Naturalist, pp. 1-4.

Spencer, K.G. 1965. *Clifford Oakes 1894-1965.* Bird Study. **12**, 261.

Spencer, K.G. 1965. *A survey of Lancashire names for birds.* Journal of the Lancashire Dialect Society, **14**, pp. 3-25.

Spencer, K.G. 1966. *Wild birds in Lancashire folklore.* Journal of the Lancashire Dialect Society, **15**, pp. 2-15.

Spencer, K.G. 1966. *Some notes on the roosting behaviour of starlings.* The Naturalist **898**, pp. 73-80.

Spencer, K.G. 1969. *East Lancashire reservoirs: an ornithological survey.* MSc Dissertation, University of Salford. Available for consultation in Burnley Library: Reference LB3/SPE and Local LB3/SPE.

Spencer, K.G. 1973. *The status and distribution of birds in Lancashire.* Privately published.

Spencer, K.G. 1975. *The house-martin at home.* Countryside, pp. 482-484.

Spencer, K.G. 1977. *The status and distribution of birds in the Burnley area.* Privately published.

Spencer, K.G. 1978. *Roosting behaviour of migrant house martins.* British Birds. **71**, p. 89.

Spencer, K.G. 1997. *Rooftop assemblies of lapwings.* E.L.O.C. Bird Report, pp. 59-60.

Spencer, K.G. 2009. *Blackbirds eating fuchsia seeds.* British Birds, **102**, p. 144.

Tate, P. 1979. *A century of bird books.* Witherby, London.

Chapter 5
Birds in Poetry
Plates 14 - 16

Ken obtained a degree of Bachelor of Arts at the University of Leeds in 1952. He was keen to take his studies further and, in 1957, he was awarded a Master of Arts with distinction for his thesis on *Birds in English Poetry from Earliest Times to the Present with Special Reference to the Period 1700-1956* (photo 14). He contended that it took an ornithologist with literary interests to judge bird-poetry adequately. In his thesis, he acknowledged a debt to Clifford Oakes who, ever since they first met in 1944, had inspired Ken with his artistic approach to ornithology, in particular conversations on bird-poets and their work.

It would be inappropriate of me to attempt a summation of Ken's thesis. Like John Clare, Ken was always delighted when an object in nature brought to mind *"an image of poetry"* associated with a favourite author. Thus he seldom saw a flock of lapwings over bleak moorland in autumn without being reminded of:

LAPWING by Rex Warner

Leaves, summer's coinage spent, golden are altogether whorled,
sent spinning, dipping, slipping, shuffled by heavy-handed wind,
shifted sideways, sifted, lifted, and in swarms made to fly,
spent sunflies, gorgeous tatters, airdrift, pinions of trees.

Pennons of the autumn wind, flying the same loose flag,
minions of the rush of air, companions of bedraggled cloud,
tattered, scattered, pell-mell, diving, with side-slip suddenly wailing,
as they scale the uneasy sky flapping the lapwing fly.

Plover, with under the tail pine-red, dead leafwealth in down displayed,
crested with glancing crests, sheeny with sea-green, mirror of movement

> of the deep sea horses plunging, restless, fretted by the whip of wind,
> tugging green tons, wet waste, lugging a mass to Labrador.
>
> See them wailing over high hill tops with hue and cry,
> like uneasy ghosts slipping in the dishevelled air,
> with ever so much of forlorn ocean and wastes of wind
> in their elbowing of the air and in their lamentable call.

Ken considered Warner's poem to be brilliant, practically every word adding to a superb mental picture, and commented:

"Warner clearly associates the lapwing with autumn, a fact readily understandable to the ornithologist, yet difficult to explain in so many words. Perhaps Warner feels the 'loneliness' of the flocks is in harmony with the rather sad season, that their wild flying and 'lamentable call' suggest the ragged clouds and whining winds of autumn days. On less romantic grounds, it would probably be true to say that there are probably more large lapwing flocks in Britain in autumn than at any other time of year.

There is a subtle comparison between the whirling autumn leaves and the plunging lapwings, nicely expressed by a thoughtful choice of words. The two subjects are similar in that they are both associated with windy days, with loneliness, with the closing year: yet they differ in that the leaves fly lightly at the whim of the wind, the lapwings heavily, pushing against the gale. _Sifted_, _lifted_, _airdrift_, convey the idea of lightness in the leaves; _scale_, _plunging_, _tugging_, _lugging_, _elbowing_ suggest the more laboured struggle of the birds.

The 'floppiness' of the lapwings' flight is neatly suggested by phrases like _pennons of the autumn wind_, _slipping in the dishevelled air_. The dreary yet not entirely unpleasant loneliness of their habitat is conveyed by _high hill tops_, _wet waste_, _forlorn ocean_, _wastes of wind_. The impression is that there is something unearthly about them: _companions of draggled cloud_, _scaling the uneasy sky_. A hint of mystery is added by the comparison to _uneasy ghosts_, to the _deep sea horses_. The phrase _lugging a mass to_

Labrador heightens the sense of fascinating speculation ('Here today, a thousand miles away tomorrow' is the idea). I have no doubt that the specific reference to Labrador has its origins in the remarkable incident of December 1927, when a large number of lapwings, one of which had been ringed at Ullswater, Cumberland, achieved notoriety by making a trans-Atlantic crossing and reaching Newfoundland and Labrador (British Birds, 22, pp. 6-13).

The birds' plumage is excellently depicted, pine-red in fact describing the colour of the under-tail coverts more graphically than the Handbook of British Birds pink-cinnamon. The burnished green of the upper parts is nicely suggested by sheeny seagreen, the deep sea and green tons. Possibly the mention of the birds being tattered alludes to their being in moult. With lapwings, moult extends from late May until November and, during that period, individuals commonly lose a few flight-feathers from each wing, which does indeed give them a very ragged appearance.

The side-slip suddenly wailing is a feature very frequently to be observed in lapwing flocks."

In 1954, whilst undertaking his dissertation, Ken accepted an appointment as an assistant in Leeds City Libraries, working at Headingly and the Central Library. In 1956 he underwent formal training at Library School in Manchester.

Whilst in Leeds he wrote a paper on the birds of Woodhouse Moor (1956). Soon afterwards, he was offered the task of cataloguing James Fisher's library (a leading natural history broadcaster / ornithologist) but declined, as he did not wish to be based in London.

References

Spencer, K.G. 1956. *The birds of Woodhouse Moor, Leeds*. The Naturalist.

Spencer, Kenrick Parker Grant. 1957. *Birds in English Poetry from Earliest Times to the Present with Special Reference to the Period 1700-1956.* Thesis submitted for the Degree of Master of Arts, University of Leeds. (Copies are held in the library of the British Trust for Ornithology at Tring and in the Burnley Community History Library.)

1. Ken at one month and Philip aged six with Mother – July 1928.
(T. Taylor).

2. Ken aged two and Philip aged eight with Mother – November 1930.
(T. Taylor).

3. Ken aged two and Philip eight with toy plane. – November 1930.
(T. Taylor)

4. Ken on 4th June 1930 with Pippin. *(Home photograph).*

5. Ken on 24th June (his 3rd birthday) 1931 – playing with Philip in field at Lower Holes Farm to rear of Landseer Close.
(Home photograph).

6. Philip and Ken about 1938 or 1939 *(studio – probably T. Taylor).*

7. Philip and Ken about 1940 – *(studio –T. Taylor)*.

8. Bus marooned in snow near *The Waggoners*, 1940. *(Philip Spencer)*.

9. Fields at Lower Holes Farm with Landseer Close in the background, taken by Philip Spencer, 15th April 1940.
(Burnley Reference Library Collection)

10. As a coal-mining optant in 1947 at the back-door of 18 St Matthew Street.

11. Ken on his 21st birthday, 1949.
(Taylor and Lambert, 105 Manchester Road, Burnley).

THE
LAPWING IN BRITAIN

Some Account of its Distribution and Behaviour, and of its Rôle in Dialect, Folk-Lore and Literature

BY

K. G. SPENCER, B.A., M.B.O.U.

Photo by]   *[A. W. P. Robertson*
A TYPICAL MALE LAPWING IN SPRING PLUMAGE

Size Demy 8vo ($8\frac{1}{2}'' \times 5\frac{1}{2}''$), 12 photographic plates, cloth boards, 16/- net. Both in the newspaper press and in the specialised journals *The Lapwing in Britain* has received excellent notices. Extracts from a few are given on the other side of this leaflet.

Obtainable through all Booksellers

A. BROWN & SONS, LIMITED
32 BROOKE STREET, HOLBORN, LONDON, E.C.1
AND AT HULL

[Please turn over

12. Flyer for *The Lapwing in Britain* published in 1953.

13. Taking part in athletics event held in a Clitheroe street about 1955.

14. Graduation Day 28th June 1957; Leeds Town Hall – *(home photograph)*.

15. Ken's mother – probably on the occasion of his graduation.

16. Ken
(Taylor and Lambert).

17. Clearing scrub from Meathop Moss with Clive Hartley on a Conservation Corps work party during 1960s *(home photograph).*

18. Reclamation at Towneley Colliery spoil heap during the 1960s I.
(Burnley Express).

19. Reclamation at Towneley Colliery spoil heap during the 1960s II.
(Burnley Express).

20. Planting bulbs with Girl Guides at Holy Trinity Church.
(Burnley Express).

21. Members of the Civic Trust planting bulbs. Front from left: Ken, Victoria Towneley, Ian Rawson, Peregrine Towneley, Simon Towneley. Back: Barbara, Tom Huggon, J. Myers.
(Burnley Express 14th October 1969).

22. With Stanley Jeeves of the *Council for the Protection of Rural England* and helpers clearing refuse near Briercliffe about 1970.

23. With Tony Audus, *Burnley Parks and Recreation Department*, and *Towneley Junior Naturalists' Club* planting snowdrop bulbs supplied by Burnley Civic Trust in Towneley Park.
(Burnley Express and News, 7th November 1972).

24. Founders of the Civic Trust: Harry Schofield, Ken and Simon Towneley. *(Burnley Express)*.

25. Civic Trust: Ken with Mary Davenport – one time teacher at the High School. *(Burnley Express)*.

26. Ken with Jane about 1979 at Healey Heights
(Burnley Express).

27. Ken conducting judge and his party through Towneley in respect of the Burnley by-laws dispute, about 1978.

28. Ken relaxing at 167 Manchester Road, undated *(photograph by Barbara)*.

29. Barbara – shortly before she retired.

30. Ken with Lawrence Chew at the Weavers' Triangle visitor centre in April 2007. *(Roger Creegan)*.

31. The author with Ken in Scott Park, Christmas Day 2009. *(Sharon Parr)*.

Chapter 6
Animal welfare
Plate 26

The Burnley Express for 1st January 1958 announced that four Special Constables with a difference had been sworn in at Reedley Magistrates Court. They were all bird-watchers whose special duty would be *"to protect birds and take prompt action against those who attempt to harm them."* The four men were: Arthur Woodward, Harold Smithies, Derek Sodo and Ken Spencer. This was exceptionally far-sighted and only recently has wildlife protection been similarly emulated on a wider scale. Their appointment had been prompted by the Bird Protection Act 1954 and the high level of indiscriminate shooting of birds around Burnley. Ken continued in this role into the early 1980s. On his walks in the countryside, he would tackle anyone with a gun and point out the laws relating to their use – a task requiring great diplomacy and at times not a little courage. In all those years, there was only one occasion when a gun owner turned awkward.

In July 1960, I joined Ken for a fortnight whilst he assisted Bob Chesney of the Nature Conservancy with the wardening of Scolt Head Island National Nature Reserve on the north Norfolk coast. On my arrival, we went straight out to the island because, whilst waiting for me in Hunstanton, Ken had found – of all things – a day-old ringed plover *Charadrius hiaticula* chick. Ken placed the chick with others which had hatched two days earlier. My notes record that *"the parents at first showed some aggression, but the female seemed to accept it, if not the male, and began brooding it."*

The skills of ornithologist and bird welfare came together in an incident reported in the *Burnley Express* of 21st November 1964, entitled *"60 birds fell down office chimney."* Starlings *Sturnus vulgaris* normally flew over Burnley to a communal roost near Haslingden. On this particular evening, in dense fog over the town, they pitched instead on office buildings in the town centre and some fell down the chimneys. Anticipating that this might happen, Ken was on hand the next morning when the offices were unlocked and assisted with the capture and release of the starlings.

Commenting on an observation by Dare & Mercer (1968) that 3.3% of coastal wintering oystercatchers *Haematopus ostralegus*, many of which were thought to be migrants from the Faroes, had injuries to their legs and feet caused by sheep's wool, Ken suggested (1968) this may be related to *"the fact that in those islands it is the custom to let the fleece grow until it begins actually to fall out."* He went on to compare it with a *"localised hazard ... in cotton-manufacturing districts where loose strands of thread form a menace to birds, especially feral pigeons."* Three pigeons brought to him in Burnley in the previous year had each lost several toes in the manner described by Dare & Mercer.

Writing in 1997, Ken's partner, Barbara Bailey said of him *"All his life he has been prepared to take in injured wild birds and care for them and there has always been a steady flow – by word of mouth. When I first became involved in the care of injured birds, I thought of it as a help to the birds. I soon realised what a social service it is. People bringing the birds to Ken care very much for the bird's welfare and feel helpless. Ken's help is even more important for children whose characters are being moulded. A defence against feeling helpless is hardening of the heart, but to share concern with someone can help to strengthen their compassion."*

Even as a boy, Ken's concern for creatures in difficulty began to manifest itself. His diary for 16th February 1941 records *"Walked through park at 11.45 with Dad and found a bird minus a leg. I got some bread from Gert's, put it in some bushes and left it there. Good luck bird [sparrow]."*

On 12th June 1941 *"Ma and Dad [went for] run in car in afternoon and brought home a frog so took him into Devil's Stewpot and let him go. He swam away."* The Devil's Stewpot was the local nickname for a deep pool in Scott Park.

Pigeons featured a lot in Ken's life. The biggest character was a female nicknamed Squeaker because, at the time she came into his life, that is what she was – an almost naked squab which had fallen from a nest at Towneley Hall. Ken bundled the young bird into his shirt to keep her warm and he and Barbara reared her to adult-hood. At that time, Ken was living at one side of Scott Park and his mother was still living on the far side in Landseer Close. Ken would walk through the park each day to visit her and Squeaker would run alongside; she would also accompany him to the shops and, rather than risk leaving her outside, Ken would put her inside his shirt until they came out, when Squeaker would walk home beside him. Squeaker turned out to be a good mother and, over the years, reared several abandoned pigeon chicks brought to Ken.

It was through Ken's kindness that Bonnie Brid came into his life. One day, looking out into Manchester Road, he saw an emaciated whippet walking towards the town. It was a Saturday and, later in the day, the same dog was looking vulnerable and uncared for at the bus station. It had come to the attention of two ladies who, concerned by its emaciated state, had bought it a meat pie. The dog, however, was too nervous to take it. Ken did not like to think of the dog on its own when the football crowds emerged. The two ladies suggested to Ken that, if he fancied a dog, the meat pie would probably be sufficient to use as a bait to tempt it home. Ken had never thought of having a dog but felt he could not abandon it so, holding the meat pie behind him at dog-nose height, he walked home followed by the dog. This dog became Brid (*"Thou art welcome Bonnie Brid"* – Samuel Laycock in Stedman 1895) and lived out her days with Ken and Barbara.

On one occasion with Brid, Ken was bird-watching and, not wanting her to disturb the birds, bade her to lie down. Shortly afterwards, he encountered the farmer and thought it polite to check he did not mind Ken being on the land. *"You're alright, provided you don't have a dog"* said the farmer. *"Well, actually, I do"* replied Ken. *"What, with you?*

Where is it?" "She's lying down over there". "Go on, then, call her up" at which Brid came bounding over. The farmer was amazed and said that, if Ken had such good control over his dog, she was welcome too.

When, after five happy years, Brid died, she was replaced by Jane – a cross whippet-grey hound – a rescue dog. Again, she proved to be a biddable companion to Ken and Barbara. Jane died in September 1986 and was succeeded by a third whippet, Judy – another much loved dog. After Judy died in 1996, Ken and Barbara felt they were too old to cope with another dog.

Ken's letters to me record many instances of assisting birds in distress:

29th April 1995: *"The Fire Brigade called this afternoon with a tiny chick mallard, rescued by a girl from two lads around a bonfire to which the brigade had been called. There seemed no point in taking it back and I said I would try to find a female with a big brood on the canal who would perhaps not notice one extra. I couldn't find one, but I did find a female with two chicks of similar size to the orphan and to my surprise she took it on without batting an eyelid."*

4th June 1999: *"I spent a lot of time and effort last week catching a pigeon on the canal bank; its feet were tangled in thread / fishing line. I managed it at last. I brought it home and Barbara and I had a full half-hour in getting it clear. Every shred needs to be carefully removed, and it's a two-person job. Very satisfying when accomplished. We reckon that we have caught and cleared something like at least 100 over the years we've been together."*

23rd October 2001: *"A curious incident last night. Barbara and I had just got settled into a Civic Society meeting at the Town Hall when the chairman came in a minute or two later and said to me 'There's a female mallard outside that's just had a collision with a bus: can you help?', so I went out and sure enough there she was in a doorway nearby. I caught her carefully in my jacket and took her up to the canal towpath. She* walked to the edge and into the water. I could hear others calling softly not far away so I'm sure she would be all right. I could see no injury, and could only assume that she had had a bang that had shaken her. What makes the incident even more remarkable is that I was called on by the Police about two months ago in an almost identical case. A female mallard had been brought in by boys who had seen her fly against a high wall; this was at night, as yesterday. She was sitting quietly in a box at the Police Station, and just as last night I took her safely up to the canal. I wonder in retrospect if it could even have been the same bird, perhaps one with a slight navigation defect."

17th July 2003: *"Did I tell you about the pigeon chick that we successfully reared to maturity? It now lives on the church but still comes down to the hand for food. It recently came into the house via the back door and went right through to the box just inside the front door where it had grown up."*

1st January 2007: *"Good news: - I told you that the pigeon 'Daisy' was missing. ... Yesterday, she turned up – a bit shattered but more or less herself. She had been away 11 days, heaven knows where. Barbara and I brought her up from being very small, 3 years ago, so you will understand that she's been special to me."*

28th September 2008: *"I was at the morning concert yesterday at St. Peter's. Coming back to the library via the pedestrian precinct I noticed a pigeon on the point of dying on the pavement. I picked it up in hopes that it might recover, but it did die. Football hooligans en route to Turf Moor had kicked it. I was just too late to see the incident, but it couldn't have been anything else."*

On principle Ken had no time for game-keeping. *"Another bad case this morning."* (Gamekeeper is jailed over snaring badgers, The Times 13th July 1999 re John Drummond of Holker Hall). Ken felt that the basic thing about game-preserving and also fox-hunting *"is that it is ethically wrong."*

References

Dare, P.J. and Mercer, A.J. 1968. *Wool causing injury to legs and feet of oystercatchers*. British Birds **61**, pp. 257-263.

Spencer, K.G. 1968. *Wool and cotton causing injuries to birds' feet*. British Birds **61**, p. 469.

Stedman, Edmund Clarence (Editor). *A Victorian anthology, 1837-1895*. Cambridge, Riverside Press 1895; New York.

Chapter 7
Historian

In 1989, as part of a natural history of east Lancashire, Ken was invited by Raymond Pickles, then in charge of Burnley Libraries, to write about the birds. Ken wondered whether to do so, *"since over the past ten or twelve years I have come to regard myself as a local historian rather than as an ornithologist."* Ken's contribution to local history since that time has been immense.

To read Ken's diary for 1941 (selected extracts - Chapter 2) is to realise how different Burnley was then compared with today. Some aspects relate to Burnley in wartime which, for a few years, with the temporary relocation from London of the *Old Vic* and *Sadlers Wells*, was able to enjoy world-class entertainment. There emerges a picture of Burnley beset by deep snow; frequent visits to the theatre, opera and cinema; radio broadcasts; a busy social scene; war-time with its barrage-balloons, 'alerts' and 'all-clears', air-raid shelters, gas attack rehearsals and bombing raids in Manchester (audible from Burnley); vegetable gardening and keeping hens – all as part of self-sufficiency; a world before emails and mobile phones in which communication was by conversation, telephone and post; a world without refrigeration and super-markets; life on the edge of town, overlooking a dairy farm where hay was still made.

Life evolves and no doubt the Burnley of sixty years hence will be very different from today. Concluding his review of his grammar school days (2003), Ken quoted L.P. Hartley in *The Go-between* who said *"the past is a foreign country"*, but Ken went on to say: *"It is; but sometimes the present is the foreign country to those of us who knew the past!"*

Dating from a walk in the hills in Rossendale, Ken took an interest in the *Larks of Dean* or the *Deighn Layrocks*, a group of handloom weavers and farmers who, in the 18th Century, were well known musicians. They sang much of Handel's music but also composed their own songs such as *"PhlymPhlam"* or *"Ca' it what tha' likes"* (Red Rose Magazine). They assembled at Dean Farm, now abandoned, to rehearse. On Sundays, they made their way across the hills to play at Goodshaw Chapel. In 1996, the *Antiques Roadshow* featured a double-bass bought in Skegness which had belonged to George Hudson. Ken commented that this would have been George Hudson of the Deighn Layrocks, originally of New Laithe Farm at Dunnockshaw; at one time the Hudsons made up the entire pier orchestra at Skegness. Ken also pointed out a connection to George Butterworth, the composer of such promise who died as a young man in the First World War and who wrote *The Banks of Green Willow.* Butterworth was directly descended from the musicians at Goodshaw.

Dean Layrocks
by Jennie Nuttall.

The History of Lumb in Rossendale.
Typescript, Rawtenstall 1965

They were mostly weaver-farmers and quarry workers. In their spare time they met at each other's houses to practice, and doubtless pass on the gossip of the day (as any chorister can vouch). At these meetings, they each paid 2^d, for which they received bread and cheese and ale brewed especially for the occasion. They were by no means rich, and could not afford to buy copies ... their 'parts' had to be copied from borrowed scores, the lines being ruled by hand; the pricking done by a feather or quill pen, which they first had to cut, split, and trim. These copies were boldly pricked and could be read at a distance of several yards.

In winter, the Layrocks had to sing and play in poor candle-light, usually tallow dips; perhaps even 'bog candles'. Space was often limited, much room being taken by the handlooms. A gentleman who looked in on one of their practices, late at night, describes them as being arranged

> according to height round a scrubbed deal table, joining at one copy. Hence the need for bold copies.
>
> Mostly they kept to Handel, but had among them prolific composers whose tunes teemed with runs and repetitions, revelling in such quaint names as 'Bocking Warp', 'Spanking Roger', 'Solemnity', 'Linnet', 'Plover', 'Skylark' and many others.
>
> Their music had several clefs. A story is told of a pupil of Robert Ashworth of Moss Barn, who was so awestruck by his teacher's familiarity with the bewildering array of clefs that he fearfully exclaimed, *"Ony mon as can teych thad, con witch! Let's ha' mi cap"*. (Any man who can teach that, is a witch! Let's have my cap.) He hurried away, and ended his musical career abruptly.

Ken wrote a number of historical booklets; checking the catalogue at Burnley Library in 2009, Susan Halstead, then Reference Librarian, found 186 items. Ken's work on local people, past and present, many of national significance, deserves wider recognition. It enthused Susan and her colleague, Diane Schofield, to produce *Famous People of Burnley*, which has now been added to the *Community History Collection*.

In 1987, Ken wrote a booklet on Tattersall Wilkinson, a 'character' of many diverse interests, author, historian, archaeologist and lecturer at the centre of a discussion group and co-founder of the *Burnley Literary and Philosophical Society*. Ken's tribute is subtitled *"a very remarkable man"*, which could equally be used by others to describe him.

For Burnley and District Historical Society, Ken wrote *An outline history of Habergham Eaves* (Spencer 1989a) and complemented this with a book of walks for the same area on behalf of the Parish Council (Spencer 1989b). More recently, he looked at some Burnley people in *They Made a Difference* (Spencer 2006), including two relatives on his mother's side. His 'Uncle' Selwyn - John Selwyn Grant - stimulated Ken's interest in Habergham Eaves, whilst James Grant, his maternal grandfather, also a historian was active from 1887 to 1890, giving talks about life in Burnley seventy years earlier (Grant 2010).

Many local history articles by Ken appeared in the *Burnley Express*. *On the Trail of the Cattle Drovers* (16th February 1982), added to an earlier feature by Titus Thornber[1]; in this he quotes from Edwin Waugh's *Besom Ben* stories relating to the early days of the previous century, a vividly-written description of drovers on the highway near Whitworth. *"They had not gone far when they came in sight of a herd of cattle, whisking their tails as they wended their way slowly onward, with two drovers lounging after them, staff in hand. In the rearward came the drovers' dogs. They had little trouble with the cattle on this quiet road; but now they darted from side to side, at their heels, and gave a bark or two, which sounded clearly in the still valley. The cattle had evidently come a long way, and occasionally one or other would stop and gaze dreamily around, and then begin to low at the sky, as if calling to the mates it had left behind among the hills and dales of the north, until roused from its reverie by a snap on the heel from one of the dogs, so that it darted forward to find shelter among the lazy-pacing herd in front. The drovers were tall, hardy men, with plaids over their shoulders; and they had a weather-stained appearance, as if they had come a long way. One of them was eating a raw turnip, from which he cut a slice now and again, with his pocket-knife."*

In the *Burnley Express* for 22nd May, 1990 there appeared *A remarkable lichen ... and its equally remarkable collector*, which included a quotation from a 1930s book *Rambles Twixt Pendle and Home* which said *"it may surprise many people to know that there is at Kew a treasured lichen from Burnley."* The specimen in question, *Usnea articulata*, is still in the Kew herbarium; it dates from the 17th Century when it was found by Thomas Willisel but it no longer grows in the vicinity of Burnley.

---

1. Titus Thornber was another well-known and respected historian who in 1987 wrote A Pennine Parish: the history of Cliviger; he was also a farmer and inventor.

Listening to Radio 4's *Something Understood* in 2001, Ken heard a rhyme which his Aunt Gert, born in 1890, used to recite to illustrate the pettiness of human squabbles. He had not heard it for over 60 years!

> *You can't play in our yard.*
> *I don't like you anymore.*
> *You'll be sorry when you see me*
> *Swinging on our garden door.*

Ken was always ready to pursue lines of research sparked by a passing interest. In the *Royal Institution Christmas Lectures 1999*, Professor Nancy Rothwell made a passing remark that as, a local historian, intrigued him. She said that up to the early years of the last century, human birthrates were perceptibly higher in March, because that was a time of year when food would be abundant. That birds time their breeding-seasons to coincide with maximum availability of food for the chicks is well-known, but the suggestion that the same concept might apply to humans was new to Ken. She said that the high March birthrate persisted up to the time when Faraday was establishing the lectures in the early 19th Century; Ken assumed her to suppose that as the industrial revolution took hold the birthrate levelled out. Spurred by this suggestion, Ken analysed the registers for St. Peter's Parish Burnley, from 1563-82 and 1633-52 but could find no clear indication of a spring peak. He also analysed the figures for Rufford, a rural parish, over the period 1800-09 but again with no significant results.

Ken took a particular interest in local people who had established a national reputation. In 2005, he researched the life of Ethel Mellor, a former mill-girl, whom he described as perhaps the most brilliant woman ever to emerge from Burnley. She was a distinguished scientist, medical researcher, linguist and lichenologist who died in 1966 aged 83.

Jacob's Join

by

Ken Spencer, 2007

As long ago as the 1940s there were inquiries in the Dalesman as to the origin of this expression. No-one came up with a solution.

A *Jacob's Join* is a social gathering to which each participant brings food to be shared with others. The *North-West Sound Archive's* publication *Sounds Gradely* (1985) recorded it in Lancashire from Sabden, Nelson, Read and Clitheroe and, indeed, this area is still the heartland of it. One of the few references to it in any other dictionary is in Eric Partridge's *Dictionary of Slang and Unconventional English* (1984) which attributed it to Mrs. Margaret Pilkington of Accrington.

A query posted on the internet on 5th August 2003 brought the response that *"it is commonly used in the mountainous area of Lancashire and Yorkshire"*, but no explanation of its origin was forthcoming.

It has variations: in Burnley it is always *Jacob's Join*, but it becomes *Jacob's Joint* round Todmorden and Saddleworth and *Faith Tea* further over into Yorkshire. At Bath, in Somerset, I am told it is simply *Bring and Share*.

Using the method made familiar to TV viewers in the recent series *Balderdash and Piffle* I have tried to work back via the local newspapers to the earliest references. These seem to be in January 1916 at St. Mary's R.C. school, Sabden (*Burnley News*, 10th January 1917). Might it have been a wartime idea designed to ease a food shortage?

My inclinations all along have been that (a) it was primarily a nonconformist 'invention' (despite the reference above) and (b) it centres on the Pendleside area[2] and (c) it started with an individual called Jacob. There is definitely no connection with the Bible.

I have a candidate for who this Jacob actually was: **Jacob Hargreaves** of Higham, a renowned local preacher (obituary *Burnley News* 23rd August 1929). He was a village postmaster and a real 'personality', very outgoing and happy to be called simply Jacob by his friends. A portrait in his memory was unveiled at *Higham Wesleyan Sunday School* in February 1930.

In her introduction to Ken's *They made a difference* (2006), Molly Haines says "When Burnley and District Historical Society was founded in 1948 one of its main aims was to promote local history. Few have done more to further that cause than Ken Spencer and most local history books written over the last few decades acknowledge his assistance with research. Ken is always most generous in sharing his sources and knowledge of the town and is always willing to help chase up elusive references ... This book is an attempt to make some of that information more widely known so that it may be used by local and family historians to further their research."

Scott Park: a memoir
by Ken Spencer, December 1996

Plate 31

I was born in 1928 at a house overlooking Scott Park, and have lived almost all of my life within a quarter-of-a-mile of it.

The park was created in 1895 on land formerly owned by the Halsteds of Hood House and farmed by Robert Schofield and his son Henry.

---

2. Susan Halstead, Burnley's Reference Librarian, has given many talks to east Lancashire groups and referred to an enquiry she had received on the origins of 'Jacob's Join'. She found that many of her audiences had not heard of the term, especially those from the Manchester area and further south.

My mother always called one part of the park 'the kitchen-garden' – because that was what it had been in the Halsted's day – and she always called the streamside 'Schofield's Clough'.

Near to the top gates, she told me, had been a little farm called Appletree Carr, which gave its name to the houses overlooking Scott Park between Carr Road and Hood House Street. Appletree Carr is inscribed on a gatepost at the top, but 'Park Avenue' is on a house about half-way down, and that became official in February 1958. The back street behind Park Avenue was known to farmers delivering milk as 'Swanks Back.'

Many's the time I've gone to sleep listening to the stream as it comes down towards the park below 'Willow Bank'. It starts at two points: one near Lower Gibfield on Crown Point road, where there used to be a trough in the wall, and one near the golf clubhouse at Glen View. For most of its early course it is (regrettably) culverted. There were four pools on the stream in the park, each dammed by a stone waterfall, and carefully de-silted every year or two. The ducks and swans were usually on the two pools below the monument. The swans were usually a pair, and there were about ten or fifteen ducks, mostly Muscovies. They very seldom wandered away; and when they did, it was a source of great excitement to me; I would chase them gently to see them fly back.

A noticeboard by the gates gave you all the park rules. "No dogs allowed except on leads" was one of them. Nameplates told you what many of the trees and bushes were; I learnt *Aucuba japonica* that way. People sometimes say that children have no sense of beauty, but I remember in detail the glorious dahlia bed that was just inside the park gates at Scott Park Road and I remember asking Mother to take a photograph of a flowering shrub that I thought was particularly beautiful. (In retrospect I realise that it was *Rhododendron praecox*.) The park was everybody's garden. My father took his cue from what was going on there – rose-pruning or whatever. The perimeter of the park was dug over every year. Some elderberry bushes, individually planted, were scattered about, and these were pruned at the same time. A few are still there, near Scott Park Road.

My best friend Brian Stuttard lived next door to us throughout my childhood. He and I called the first pool where the stream enters the park 'The Devil's Stewpot'. 'The Scott Park Ghost' was sometimes mentioned. It was all before my time, but yes, there really was such a thing; or, at least, there was once a young man going round the park at night, dressed up in a white sheet. I remember reading that he was caught and actually served a short prison sentence.

The 'park bell' was a sound that we all knew. The bell was attached to the foreman's house and rung each evening 'between the dark and the daylight' to give people time to leave the park before the gates were shut. Sometimes one of the keepers would also go round shouting "All out". After the bell was discontinued, the keeper blew a whistle as he made his way round to the gates. The park keepers were two elderly men. They wouldn't have lasted ten minutes today, but they carried a fair amount of authority then. There were two playgrounds when I was growing up; they were just open lawns. Notices said "Please keep off the grass" on all the other lawns, and you were certainly not allowed to trespass into shrubberies or down by the stream. There was no fixed playground equipment until 1977, when it was installed near the bandstand, where it is now. The park-keepers (called 'Parkies') were superseded by rangers, who were at first called 'Yogis' (after a cartoon character called Yogi Bear). They had motorcycles at first, then vans.

I was fortunate to grow up when and where I did. We had the park at the front of the house and the farm at the back. From these beginnings grew my lifetime's interest in bird-study.

Latterly, Ken took an interest in social history. He came across an entry in the *Burnley Express* for 17th July 1880 which reported *"the first case of self-murder which has taken place in or near Burnley for 20 years."* Even allowing for some suicides at that time being hushed-up because it was a Christian sin, and to attempt it was a crime in law, it seemed evident that they were rare. Despite the horrors of poverty, child mortality, disease and the 'dark Satanic mills', those decades seem to have been an age of optimism when new churches and schools were being built, sport was becoming popular, societies were being formed. By contrast, the current times appear to be an age of pessimism with churches, schools, libraries and post offices closing. Suicides in Burnley are a weekly phenomenon, often by young people who have fallen victim to drugs or debt. Ken felt that despite today's apparent higher standard of living, people overall seem less fulfilled – a sad reflection on modern society.

Ken's knowledge of Burnley's history and of the library's Local Studies collections was greatly valued by the professional staff. In December 2007, Ken happened to see an email from Diane Schofield, one of the librarians, in response to an enquiry in which she said: *"Ken is quite brilliant regarding the scope of the Local Collection here at Burnley."* Ken commented *"Barbara would have been pleased. I am, I'll admit"*.

References

Grant, J. 2010 reprint. *Regency Burnley: a tour of the town in 1818. Burnley & District Historical Society. Comprises: (1) A sketch of Burnley 70 years ago. (2) Burnley 70 years ago: its by-ways.* Papers read to the Burnley Literary & Scientific Club 1887 & 1890, respectively.

Haines, M. 1998. *Not just another Corporation Committee: 50 years of the Burnley and District Historical Society.* Retrospect: Journal of the Burnley and District Historical Society. **16**, pp. 3-9.

Hall, B. and Spencer, K. 1993. *Burnley: a pictorial history.* Phillimore, Chichester.

Nuttall, J. 1965. *The history of Lumb in Rossendale*. Typescript, Rawtenstall Public Library.

Pickles, R. (Editor). 1989. *The countryside around us: a natural history of east Lancashire*; includes a chapter on *The commoner birds* by Spencer, K.G. Lancashire County Council Library, Museum and Arts Committee.

Red Rose Magazine Dec 1989/Jan 1990. *"PhlymPhlam" or "Ca' it what tha' likes" - some thoughts on the names of tunes composed by the Larks of Dean*.

Spencer, K.G. 1987. *Tattersall Wilkinson: a very remarkable man*. The Briercliffe Society.

Spencer, K.G. 1989a. *An outline history of Habergham Eaves*. Burnley and District Historical Society.

Spencer, K.G. 1989b. *Walks around Habergham Eaves*. Habergham Eaves Parish Council.

Spencer, K.G. 1997. *The houses round the church*. Outreach: the Parish of St. Matthew the Apostle and Holy Trinity, Habergham Eaves.

Spencer, K.G. 2000. *Sunny Bank School: some notes and memories*. Retrospect Vol. **18**, pp. 20-24.

Spencer, K.G. 2003. *Burnley Grammar School 1939-46: some notes and memories*. Burnley and District Historical Society.

Spencer, K.G. 2004. *Burnley Parish Church of St. Peter. Inside the church: a look at some of the memorials*. Leaflet.

Spencer, Ken. 2006. *They made a difference: notes on some Burnley people*. Burnley and District Historical Society.

Chapter 8
The By-Laws Dispute
Plate 27

Mr. D.K. Hall, Editor of the *Burnley Express* at the time of the Burnley By-Laws saga, which ran for eight years from 1976 to 1984, described it as *"... possibly the biggest controversy our town has ever known."* Ken was a member of the *Burnley Dog Owners' Action Committee*. He felt it important that *"someone who really knew the story"* should write an account. As he saw it, *"the Burnley Battle was not primarily about dogs, or dogs in parks. It was about by-laws, the way they are made and subsequently enforced. Local Government, in other words – something that concerns us all."* (Spencer 1989).

By-law 15 first came to Ken's notice in 1977, with the posting of notices announcing that, henceforth, dogs could not be taken into Scott Park. It was claimed by the Borough Council that the by-laws had been fully debated, and approved, the previous year by the Home Office. An informal action group was formed in the hope of reaching an amicable solution. The strength of feeling became evident when hundreds of people and their dogs turned out for a demonstration at Scott Park.

At a stormy meeting convened at the *Sparrow Hawk Hotel* and attended by several councillors, it was resolved to form a committee to represent Burnley dog owners; Derek Baker, Frank Clifford, Colin England, Harry Baxter, George Hughes, Mavis Thornton and Ken were elected. A subsequent proposal that the by-laws be suspended was defeated and a second protest was announced for Monday 15th August. On the evening of Friday 12th August, however, the members of the Burnley Dog Owners' Action Committee were served with summonses requiring them to attend the High Court in London on the Monday. This had the effect of escalating what had been a local dispute into one under the national spotlight.

At the High Court, Mr. Brian Leveson, Q.C. for the Council, described the defendants as *"the kind of people against whom Society needs to be defended."* Mr. Justice Slade imposed an injunction on them ordering that, pending a full trial at which its validity could be challenged, the by-law must be observed.

In the face of escalating costs, four members of the Action Committee stood down, leaving Frank Clifford, Mavis Thornton and Ken to go it alone. Attempts by the *Burnley Express* to mediate went unanswered by the Council. Some went to prison and others were fined for contravening By-law 15. The Council claimed that *"At all stages in the making of the by-law the matter was widely debated in Committee and Council and was fully reported by the news media. All discussions took place with the press and public present."* Whilst the public notice relating to the by-laws had appeared in the local papers on 16th November 1976, it simply stated that the Council had applied to the Home Office for confirmation of new by-laws affecting its 'pleasure grounds', that a copy of these by-laws could be seen at the Town Hall and that any objections could be made to the Under-Secretary of State. It made no mention of dogs.

The origin of the demand for tighter by-laws could be traced to two councillors from Burnley having attended an environmental health conference in 1975 at which Professor A.W. Woodruff had spoken on *toxocariasis* – a parasitic infection spread via dog faeces and which, if ingested by a young child, can cause blindness.

The trial upheld the by-law and costs of approximately £13,500 were awarded to the Council, leaving the defendants to raise the necessary money. On 19th July, Frank Clifford, taking a deliberate and carefully-considered decision, walked into Scott Park with his dog. This led to his re-appearance before the High Court in London. He avoided prison by giving an undertaking not to infringe the by-laws in future.

On 20th February 1979, Mavis Thornton entered Thompson Park with her dogs and continued to do so until, on 9th March, she was summoned to the High Court and sentenced to three weeks in prison. On her release, Mrs. Thornton said she had not gone to prison on behalf of dogs, but on behalf of people. The Council continued to summon others for breaking By-law 15 and further prison sentences followed.

A further attempt by the Bishop of Burnley to negotiate a compromise was rejected by the Council. The local MP, Dan Jones, became concerned that the by-law was giving Burnley a bad name. *"It is becoming a standing joke in the Commons and the serious concern is that the publicity could affect the district economically through developers showing reluctance to invest in a move to a town which seems unable to resolve such an apparently trivial issue."*

At the time the costs were awarded against the three defendants, there was perhaps no expectation by the courts that Burnley Council would pursue them. When, however, it became evident the Council was determined to do so, public sympathy identified increasingly with the defendants. The *Sunday Express* considered the case absolutely monstrous and the only winners *"the sleek lawyers laughing all the way to their banks."* Before the Council could lodge its claim, the High Court Taxing Officer had to examine it and in doing so reduced the demand by £5,000. In the interim, public donations enabled the Action Committee to pay its own legal costs.

Ken's personal 'share' of the Burnley Council bill came to £3,605. In an open letter to the council's Chief Executive dated 10th July 1980, Ken made it clear he felt no moral obligation to pay. The Council resolved to instigate a means-test against the defendants and appeared to have their sights set on the Dog Fund. This fund, however, was not available to the defendants. At the examination, the Council had failed to ask that the costs of holding it be awarded against the defendants and themselves began to incur costs amounting to £1,250. Despite Councillor Pike, the

Evening Star and the council's Chief Executive Officer advising against it, the Council now voted to commence bankruptcy proceedings against the three defendants. When it became evident that this would lead nowhere, the Council resolved to suspend such proceedings *"until such times as the financial circumstances of one or all of the defendants might change for bankruptcy proceedings to be worthwhile."*

As a consequence of the *Local Government Act 1972* and the *County of Lancashire Bill*, all local legislation in the shire counties was to cease to have effect by the end of 1984. One of its purposes was to get rid of archaic Acts like the *Burnley Borough Improvement Acts* on which Burnley's by-laws were based. Burnley Council had wanted to retain the by-law by means of a 'saving clause'. However, with the publicity which now surrounded the case, this proved contentious and a hearing before a Select Committee cost the Council, i.e. the ratepayers, £6,698. Again, suggestions that Burnley Council might abandon its stance and re-apply to the Home Office in the orthodox way came to nothing. With others, Ken petitioned the Council to show proof-of-need for its dog-ban by-law.

Petitioning is not to be undertaken lightly and a hearing before the Select Committee is very like a court case. Ken and Barbara, who represented themselves, were grateful for help on procedural matters to Lord Houghton who had taken a personal interest in the case and to *Lancashire County Council*, the *Home Office* and the *House of Lords Private Bill Office*. The hearing lasted seven days. The outcome was that Scott Park was exempted from the by-law until such time as it had been restored to pristine condition – a victory in part.

The whole dispute could have been settled round a table in the Town Hall. Ken reflected: *"It was an appalling reflection on the town's management that a decision which should have been its own had to be taken at the end of the day by a committee in the House of Lords."* Even then, it was not the end of the saga; the Council had to be instructed not to continue upholding By-law 15 pending enactment of the *Council of Lancashire Bill*.

References

Spencer, K.G. 1989. *The Burnley By-Laws Dispute*. Privately published.

Chapter 9
Foot and Mouth Disease

The way in which the outbreak of Foot & Mouth Disease (FMD) was handled in 2001 resulted in the year being a write-off for anyone wishing to pursue field work. Ken was particularly frustrated, commenting that: *"It seems to me that the western world has gone mad over this."* In a letter of 20th March to Burnley MP Peter Pike, Ken asked *"How, by going on a circular walk from this house over the moors, keeping to public footpaths, would I in any way be endangering the sheep and cattle?"*

Ken agreed with Simon Jenkins who wrote in *The Times* of 14th March 2001: *"The Government has gone mad. It is reportedly hiring snipers to shoot wild deer on Dartmoor and wild boar in Essex to protect the incomes of a small group of businessmen. The plan of the Agriculture Minister, Nick Brown, to use the Army to save his 'national herd' is ludicrous. His panic reaction to FMD is disproportionate to any conceivable public harm. It shows how easily modern government is corrupted by a commercial lobby ... Let us say it again. The FMD epidemic has nothing to do with public health. It is to do with money. The disease does not kill people or make them ill. To animals it is distressing but not fatal and was traditionally treated with tarred rags until the victim recovered ... In the case of F.M.D. I see nothing but a bureaucracy struggling to safeguard an already protected industry ... a cost to tourism of some £250 million a week is considered a price worth paying against £20 million a week to farmers."*

An inn at the summit of Manchester Road which had for long been known as *The Bull and Butcher* had chosen to rename itself *The Slaughtered Lamb*. However, in the light of FMD the landlord decided to revert to the original name – perhaps, Ken observed, the only good outcome of the whole sad episode.

Ken was struck by the inconsistencies of the FMD restrictions. Hence, while he could not walk round Clowbridge reservoir or the track through the fields over Crown Point, there would be nothing to prevent him from walking from Padiham to Clitheroe over the Nick-o'-Pendle, where sheep constantly wander on and off the road.

Writing a few days later in *The Times*, Magnus Linklater wondered why ministers had not sought advice from the real expert, Professor Fred Brown (who, Ken pointed out, was an old boy of Burnley Grammar School). Magnus Linklater described him as probably the world's leading expert on FMD and went on to report that: *"On 9th March, four days before the Ministry of Agriculture announced its slaughter policy, he offered to supply a system that his laboratory had developed, which could test for FMD infection on site and produce accurate results within two hours. For sheep, in particular, where symptoms are hard to detect, it would indicate straightaway whether a flock was infected or not. The slaughter of healthy animals to create a disease-free barrier would be unnecessary. The testing kit was simple, cheap and effective. By any standards, it was a vital breakthrough. Used properly, it would have avoided the funeral pyres and stinking piles of unburied animals that litter the countryside. The scientists at MAFF, however, were not interested. They were 'too busy",* they said. *"That saddened me a lot,"* said the professor ... *"Why destroy innocent healthy animals? ... The policy was being pursued because MAFF concluded, wrongly, that it was the only way to stop the spread of the disease and protect exports. The ministry had made the crucial error of equating the present outbreak with the last one in 1967-8, which mostly affected cattle and pigs. In this case, sheep were the principal victims, and could carry the virus for several weeks before showing symptoms. The disease was therefore out of control almost as soon as it was identified. Culling animals on contiguous farms achieved nothing but unnecessary suffering. The only way of slowing it would have been vaccination. Swiftly used, that would have brought the disease under control without any burning or burying. However persuasive the argument, ministers* seemed incapable of advancing it in the face of intransigent opposition from farmers' unions and the food industry. And yet," as Professor Brown explained, "their resistance is based on outdated science and prejudice."

In July, Ken reported that: *"Peter Pike has been helpful and has sent me the page from Hansard wherein Margaret Beckett **concedes** that walkers are not a threat, which has been my contention all along."* (Hansard p.766 Oral Answers: 28th June 2001 – Paddy Tipping to Margaret Beckett re economic value of walkers as revealed by FMD crisis). Further, Ken received a letter dated 5th July from Mary Lynch, Chief Executive of the English Tourism Council, in which she said: *"I have been attending all of the meetings of the government convened Rural Taskforce ... For the last six weeks, the information given to the Taskforce is that veterinary advice confirms that there is no known case of FMD being transmitted by walkers or visitors."* She went on to say that her council predicted losses to domestic tourism for the year of £5.2bn - approximately four times the size of agriculture in economic terms.

Even when this evidence began to circulate and the Government gave the impression that the countryside was once more open to walkers, Ken and Gerry Almond found Barley to be a complete no-go area. Even in August, it was quite clear that some farmers were slow in re-opening paths, with many around Pendle remaining closed.

In Rossendale, Ken's long-time friend, Eric Ward was similarly frustrated; lack of access to the hills had restricted his activities. *"I didn't go over The Hill for the first time in more than 50 years – didn't see a lapwing on territory or hear a skylark sing – something I would never have believed possible. Yesterday, I managed a short walk over the spur of The Hill which extends towards Waterfoot and was lucky enough to see a flock of about 120 twites. I was watching them perched along a wire when they took flight as one and I looked up to see a sparrowhawk flying over. It made up for the many blank days."*

Ken was interested to learn in October *"that Elliot Morley, who is Margaret Beckett's deputy, is an ornithologist (ex-council BTO and RSPB)"* and wished he had known earlier as he would have tried a personal approach. It was Elliot Morley who relayed the message, via Peter Pike, that there had been no case of FMD attributable to an ordinary walker on a country path. In November, Ken wrote to Elliot Morley asking *"Has the Government a better strategy for the future should FMD recur? ... please, don't let us have to go through this charade again."*

Chapter 10
Photochemical smog

A phenomenon Ken consistently commented upon, for more than two decades, was that of photochemical smog. In August 1996 he observed: *"The town is blue-grey with car-gas."* In June 1999 he was so disenchanted with the weather that he wrote to the Meteorological Office to seek confirmation that there are more high winds than there used to be. Derek Hardy replied, *"I rather share your concern about the decline in numbers of certain species of animals, insects and birds, but think that one will have to look elsewhere other than the weather for the reason."* Ken wondered whether photochemical smog may be a cause of decline. Others evidently were thinking along similar lines; in a letter to *British Birds* in May 2000, Dr. Summers-Smith suggested photochemical smog might be a factor in the decline of house-sparrows.

Ken made further enquiries, regarding climate change, with the Meteorological Office in August 2002. *"My impression – short-term of course and therefore only an impression – is that we are indeed getting milder winters but wetter and windier summers."* The Meteorological Office responded: *"The main reason for the unsettled weather affecting Britain for a good deal of the summer this year is that Atlantic depressions have tended to pass closer to the U.K. than is generally the case ... because of this, the northwest of the British Isles may have experienced more wind than usual but this is not necessarily part of an ongoing trend."*

Back in the 1960s and 1970s and probably into the 1980s, in the course of a summer car journey, the windscreen would become covered with smashed insects. Their bodies contained fats which made it very difficult to retain clear visibility, such that one might have to stop en route to clean the windscreen. This phenomenon is now a thing of the past. To try and get some measure of regional variation, in 2003 the RSPB devised a simple way of sampling insect kills known as the 'splatometer'. Ken commented, *"To my mind, as you know, the suspected cause of insect declines must be photochemical smog ... what else coincides in time so exactly?"* This smog would appear to be attributable to vehicle exhausts; this was suggested in the article *Filthy Air. Pollution levels soar by half: traffic is to blame* by David Ottewell, *Manchester Evening News* 18th July 2003. In a letter of August 2003, The Friends of the Earth agreed that summertime smog *"damages plants and inhibits respiratory systems ... levels are often highest in rural, not urban areas."*

The *Lancashire Evening Telegraph* of 27th August 2003 reported *Pollution levels too high at five hot-spots*. It stated that, under Lancashire County Council's Local Transport Plan, *Air Quality Management Areas* are likely to be set up in parts of Burnley in 2004, primarily near urban roads.

2004 was a great year for privet blossom, yet poor in butterflies. Ken was sure that photochemical smog, very obviously present over the area during late July / early August, was having an effect. *"What happens when rain brings it down onto the grass?"* Bird numbers were also much reduced: *"I crossed Scott Park yesterday (3rd August); practically birdless. Some might say I'm imagining better days in the past, but I have my notes here that show this isn't so. What else is there that so exactly coincides in time with bird and butterfly declines?"*

At Worsthorne with John Slockett in April 2005, Ken was appalled by the absence of birdlife. *"When you and I were up there a few years ago there was still something to see and hear but on Friday it was totally lifeless though the higher fields look as they always did. All the air over Burnley was saturated with photochemical smog and I do wonder what effect this has. Politicians play it down of course, because of the car and aircraft industries."*

Again in late January 2006: *"Photochemical smog has been obvious these past few days. TV describes it as mist, which it certainly isn't."* In February 2006, Ken was especially interested in an item I had sent him about hedgehog decline. *"It doesn't surprise me and, as you may guess, I come back to my old bogey, photochemical smog, affecting invertebrates in the grass, compare yellow wagtail, now completely gone from E. Lancashire."*

The trend continued in April 2007 with *"garden birds ... numerically down. All that, plus the decline of pipits, skylarks, etc. coincides with the rising presence of photochemical smog, but it's not even mentioned as a possible cause. Traffic out here at 167* [Manchester Road, Burnley, BB11 4HR] *is horrendous. It must be ten times what it was when Barbara first came here about 1960."*

In early May 2008, Pendle and Longridge Fell were at times obscured by photochemical smog. Ken considered *"the Government is playing down the effects of this."* In late May *"the whole of the Ribble plain was overhung with photochemical smog. You never hear this mentioned on British TV weather programmes."*

Some years ago, *The Independent* offered a cash-prize of £5,000 for a paper published in a peer-reviewed scientific journal explaining, in the opinion of the referees, the massive decline of house sparrow populations in towns and cities. In October 2014, I wrote to Michael McCarthy, Environment Editor 1998-2013, to enquire the outcome. He replied *"In 2005, Kate Vincent, a doctoral researcher at De Montfort University, Leicester, made the discovery that sparrow chicks were starving to death in the second brood for lack of invertebrate food, and this would explain the decline. In 2008 she entered her findings, written up with other researchers from the RSPB and English Nature, for the prize, but the referees (the RSPB, the BTO and Dr Denis Summers-Smith) were split. The problem was that Kate's research explained half the mystery - but it did not explain how the lack of invertebrates had come about. One referee said: award the prize. One said: do not award the prize. The third said: award half the prize. In the circumstances, it did not seem possible to award it."* This was unfair; Kate Vincent's research had provided an explanation of the massive decline in house sparrow populations. Explaining the lack of invertebrates was an entirely separate problem which is yet to be resolved.

A note in *British Birds* by Summers-Smith (2009) suggested the introduction of unleaded petrol might be a factor. He based this on correlating its introduction in 1989 with the start of the urban decline. Whilst the evidence is circumstantial, there are indications that photochemical smog – which must get absorbed when it rains and deposited on the ground – may be the cause of the declines in insects as an essential part of the food-chain for many species. Katherine Richardson, Chair of the Scientific Steering Committee of the Copenhagen Climate Congress, commented in *The Guardian* of 11th March 2009 that even the oceans are being affected.

It would appear that Ken's observations relating to photochemical smog may well contain the answer to declining populations of once common birds and insects. Certainly, in the hills around Burnley, where the landscape remains much as it was when Ken undertook his research on lapwings in the late 1940s / early 1950s, it is not simply a question of habitat change.

Reference

Summers-Smith, D. 2009. *House sparrows and unleaded petrol*. British Birds **102**, p. 103.

Chapter 11
Civic Pride
Plates 18-25 and 30

The *Burnley Civic Society* is at the forefront of the town's civic pride; it began in 1964 as the *Burnley and District Civic Trust*, a title reflecting its work beyond Burnley, 'trust' implying dedication and care. Ken was a founder member.

An early theme was 'greening the town'. A colliery slagheap marred the main road adjacent to Towneley Hall. Ken organised its grassing and afforestation, aided by local scouts and guides and also a group of Borstal boys from Dobroyd Castle. With the aid of friends and volunteers, this was achieved for the amazingly low cost of £20, from the *Community Council of Lancashire*. Ken also supervised children in planting daffodil and snowdrop bulbs around the town. Barbara Bailey commented *"the majority of people thought it a waste of time; for a short time they were picked and even pulled up, but this did not last long and the work of those children has brightened the town for many years."*

To show that *"our town is somewhere that's really worth caring for"*, the Trust drew attention to its architectural heritage with a town trail (Catlow et al, 1983). This illuminated the history behind place names such as Bull Street, 'the Big Window' and Coal Street; it explained why the River Calder runs through a stone channel; and drew attention to such fine buildings as the Town Hall, the Mechanics' Institute and the building which now houses the *National Westminster Bank Training Centre*.

Burnley's industrial past is evident from the surviving mills, especially their chimneys, of which only about twenty remain but which once numbered 200. The town trail led onto the towpath of the Leeds and Liverpool Canal and through the *Weavers' Triangle*; the booklet expresses an aspiration *"to see this very important area of industrial architecture preserved."*

Hall and Frost (1997) wrote a fine town trail focused specifically on the *Weavers' Triangle*, with the Visitor Centre at its heart. This centre is now in good care and has been the focal point in stimulating renovation of the surrounding textile buildings. Bill Bryson signed the visitors' book whilst highlighting the possibility of *"a thrilling new chapter for the town"* (*Lancashire Evening Telegraph* 17th May 2006). On summer afternoons, Ken gets a great deal of pleasure greeting visitors at the *Weavers' Triangle*.

At the Trust's celebration of its 21st birthday at Gawthorpe Hall on 5th March 1985, Ken gave a talk, *"The First Twenty-One Years."* Richard Catlow (1985) marked the anniversary with a review in the Trust's newsletter: *"Some of our earliest achievements seem commonplace today, but we were among the pioneers of planting large trees in public areas – something that is taken for granted nowadays. Our ideas, together with not a little hard work, on ways of reclaiming colliery waste tips have been so well learnt that virtually none survive in the borough today. Similarly, it is hard to believe that advocating stone-cleaning of old buildings was ever considered revolutionary"*.

"These are general successes ... A more local success was in promoting community use and appreciation of the canal. A decade or so ago, the Leeds and Liverpool Canal which snakes through the town was unreachable and virtually unused. You had to risk life and limb scrambling over barricades of corrugated iron and barbed wire to get to the towpath. The waterway was generally considered a deathtrap ... and would have been best filled-in. Our campaigning, which involved taking the entire council on a canal cruise and years of persuasion and publicity, resulted in a dramatic change. We built a new access to the canal ourselves, through our daughter organisation Canal 2000 ... and got old access points re-opened. Now the towpath is widely used and the whole canal-side area is one of Burnley's greatest attractions, the subject of films, paintings and festivities." Just how successful this has been is evident from the 'canal event' in 2000 (see: *Candle spectacular to celebrate waterways heritage,*

Lancashire Evening Telegraph 25th September 2000), described by Ken as one of the best things ever to have been staged in Burnley. There is now an annual Burnley Canal Festival.

Catlow concluded: *"One of the main jobs of a civic society is to make people aware ... we have been very successful in this, whether it be getting the creation of conservation areas, saving fine cobbled streets from the bleak fate of tarmac, increasing the number of listed buildings or persuading the owners of character buildings in a poor state of repair that demolition is not the only answer."*

When, in 1996, Burnley won the Green Apple Award for being the most environmentally-minded local authority in England, Ken gave much of the credit to Sue McNulty of *Burnley's Environmental Forum* and *Wildlife Advisory Group*, on both of which the *East Lancashire Ornithologists' Club* was represented. Ken was also full of praise for the family events organised by the Burnley Park Rangers, which attracted a sizeable following of both young and old.

Having heard of a proposal to demolish the gatehouse to Towneley Hall and its associated arch, both listed buildings, Ken and Barbara solicited the help of local councilor Mrs Ada Smith; their efforts resulted in these buildings being saved. They also succeeded, in consultation with Alderman Gallagher and with Stevenson Memorials, in retaining *in situ* the gravestones at *St. Peter's Parish Church*. The threat of their being laid horizontally stimulated the compilation of the inscriptions on all the gravestones and memorials; these are held by the *Local Family History Society*, with copies in St. Peter's. Given the modern interest in ancestry, they provide a valued resource for those tracing family trees.

A current recruiting leaflet summarises the achievements, including publishing guides to Burnley town centre, the *Weavers' Triangle* and Padiham. The Trust pioneered the call for bottle banks and other forms of recycling in Burnley.

The 'blue plaque' scheme which identifies buildings where notable citizens lived is also to their credit.

- A plaque at 25 Scott Park Road marks where Sir Ian McKellen was born. At times, this work has its comical side, since Sir Ian declared that the plaque had been put on the wrong house, only for him to be proved wrong when Ken traced Sir Ian's birth certificate confirming that he was indeed born at that address!

- The plaque for Lee Thistlethwaite of 88 Albion Street also benefited from Ken's research. He was a cotton manufacturer, but also an accomplished musician who performed regularly with the Hallé Orchestra, both on the cor anglais and the bass oboe. In the 1920s he formed the *Burnley Municipal Orchestra*.

- A plaque has been erected in Worsthorne to Eric Halsall who promoted sheep-dog trialling, including compéring One man and his dog.

- Another has been placed at 39 Scott Park Road where the soldier Victor Smith lived. Victor died at Gallipoli on 22nd December 1915. He threw himself on a grenade to save his comrades, for which he was awarded the Victoria Cross.

All his life, Ken has taken pride in his surroundings. Fly-tipping is an incessant problem – one which the late Stanley Jeeves of the *Council for the Protection of Rural England* did much to tackle in the 1970s. A press photograph from that time shows Stanley with Ken and a group of teenage volunteers clearing a road-side near Brierclife. Even in his nineties, Ken would still pick up items – particularly those which might pose a hazard to birds such as the plastic mesh which retains drink-cans for ease of carrying, but also cans and bottles. An nonagenarian's civic pride shamed many younger people who disfigured the streets with items thoughtlessly tossed aside.

References

Catlow, Richard, Ellis, David J., and Spencer, Ken. 1983. *Burnley: a town trail*. Burnley and District Civic Trust.

Catlow, Richard. 1985. *Burnley and District Civic Trust: 21st Anniversary.* Contact: Newsletter for the Civic Trust for the North West, Autumn 1985, p. 42.

Hall, B. and Frost, R. 1997. *The Weavers' Triangle: a town trail*. Burnley Borough Council in association with The Weavers' Triangle Trust and Burnley and District Historical Society.

Chapter 12
Enthusing Others
Plate 17

Through the *East Lancashire Ornithologists' Club* and the *RSPB's Young Ornithologists*, Ken mentored young birdwatchers during the 1950s/60s, including myself. This has been widely acknowledged. Thus Fitter (1959, p.155) referred to Ken's *"gift for working with young people"*. Attending the London Olympic Games 1948, as a spectator, Ken took time to visit nearby reservoirs to watch little ringed plovers and there he encountered Richard Fitter, who gave him *"a lift back to base in his snazzy little car. He was an impressive driver!"*

On behalf of the *Conservation Corps* (predecessor to *The Conservation Volunteers*), Ken assisted Brigadier Armstrong with supervising work on *Meathop Moss* and *Minsmere Nature Reserves*. The former is a raised-mire at the head of Morecambe Bay which had been adversely affected by drainage such that the fringes were becoming colonised by pine, birch and rhododendron. Not being native, rhododendron is damaging to the entomological interest for which the mire is renowned; in 1960, under Ken's supervision, we uprooted and burnt a substantial number of bushes, no mean feat given the rainfall that Easter.

In July 2003, Ken made a return visit to Meathop Moss with Peter Hornby; he wanted to see large heath *Coenonympha tullia* butterflies. Not only did he succeed, but he enjoyed recalling plant names he had first learnt there, including bog asphodel, bog myrtle and cranberry. Under the management of the *Cumbria Wildlife Trust "the open area is now much more extensive than it was, and a lot of hydrology work is going on."*

Minsmere is now a flagship reserve managed by the RSPB; in 1961, guided by Bert Axell, the warden, we helped to build an island surrounded by water. In subsequent years this attracted nesting terns and avocets (Axell

1977). A television programme in February 1997 featured Minsmere, from which Ken was reassured to see 'our' island still providing a nesting place for terns.

In 1960, Ken assisted Bob Chesney, the Nature Conservancy's warden at *Scolt Head National Nature Reserve*; I visited him staying with Bob and his wife Phyllis at Dial House in Brancaster Staithe. In 1999, I sent Ken a *Guardian Country Diary* referring to fox predation on the NNR; Ken commented *"That wouldn't have happened in Bob's day. I remember that a stoat came onto the island; I'd seen no sign of it, but Bob had. He called it up by that call you can make on the back of your hand, and shot it dead instantaneously; regrettable but necessary."*

Resulting from a talk Ken gave to the Rotary Club, a nature reserve was developed at Towneley Hall.

Ken and I continued to visit former haunts. One, which has altered little in its overall appearance in recent years is Hyles Moor, above Bolton-by-Bowland, which once reverberated to the songs of curlews and where, even today, a few are still to be heard. Ken wrote in 2006 to Tim Melling, then of the RSPB, advocating conservation measures, who replied that he was aware of Hyles Moor: *"it is pretty good for short-eared owls and hen harriers in winter. I will mention this to my colleague Gary Woodburn who is seconded from RSPB into the Bowland Initiative ... an example of his influence is Chipping Moss ... 3 years ago this area was a jungle of soft rush with no breeding waders. Rush control, water-level management and scrape creation has resulted in lapwing, snipe, curlew and redshank all breeding on site in reasonable numbers ... A farmer near Cant Clough ... has agreed to manage his land to help twite."* This is an enormous boost to the efforts of local advocates including Ken.

In 2007, at *Burnley Central Library*, Ken got into conversation with Adele Earnshaw, a student from the University of Nottingham. She was preparing a dissertation on the history of the *Lancashire Naturalists' Trust*.

Ken provided her with a copy of part of *Nature in Lancashire*, Vol. **1**, 1970, listing both our names among its members. At its inauguration, it had no employees, compared to 92 in 2007!

Commenting on *The Birds of Lancashire and North Merseyside* (White et al 2008), Ken said *"It's absolutely superb. I'm glad to have lived to see it, and have told the authors so"*. Reviewing the history of ornithology, Malcolm Greenhalgh said *"In east Lancashire Ken Spencer, English teacher and archivist, made a detailed study of the lapwing, culminating in his book The Lapwing in Britain (1953), and took over as Lancashire recorder when Oakes retired in 1957. Many young ornithologists, the present writer included, received much help and quiet advice from Spencer, whose knowledge of wildlife and its literature remains vast."*

References

Axell, Herbert. 1977. *Minsmere: portrait of a bird reserve*. Hutchinson.

Fitter, R. 1969. *Wildlife in Britain*. Pelican Original.

White, S., McCarthy, B., & Jones, M. (Editors). 2008. *The Birds of Lancashire and North Merseyside*. Lancashire & Cheshire Fauna Society in association with Hobby Publications, Southport.

Chapter 13
'Star girl at evening'

That Ken's brother Philip was a poet is, by now, widely acknowledged. Writing poetry is intensely personal and, although I knew Ken for 50 years, it was only in 2008 that I discovered that he was himself a poet.

February 1954:

The Chiffchaff's Song
The chiffchaff's song
Brings memories of celandines
And wood anemones
Of primrose banks, and lanes in early spring.

The brook on bright March mornings
I recall: a haunt of wagtails.
Great golden kingcups blossom there
And underneath the bridge
Reflection weaves its webs of light,
Restfully restless,
In its water-play.

Autumn 1954:

If You Have Two Loaves, Sell One and Buy a Lily
Oh, lovely lady, on this day of Spring,
Why now reject the skylark's song,
Ignore the chaffinch on the blossom bough?
Why close the door on beauty
And remain with those
Who never question why they live?

Oh, lovely lady, on this day of Spring,
Come now and claim the precious things of life,

> *... Or would you never know the thrill*
> *Of holding someone close within your arms,*
> *Possess the gift*
> *Of making one man happy with your kiss?*

May 1955:

O To Be Always in the Sun!

> *O to be always in the sun!*
> *The blackbird's song from summer trees*
> *Floats leisurely.*
> *Now is the time to think of wellsprings,*
> *Ferns and dripping moss.*
> *Of shadow-valleys, mountain streams,*
> *Of iris-fields, the sea.*
> *To dream of islands: from the crag*
> *And, far beneath*
> *The swirl of water in a dark ravine.*

May 1958:

Margaret

> *Reminded of your name*
> *I once again look up to apple boughs;*
> *And hear sweet goldfinch voices*
> *In among green leaves.*
>
> *I see a beehive in the sun;*
> *Take deep delight in blue forget-me-nots;*
> *And all along the orchard walk*
> *Find pleasure in your gentle company.*

Remembering schooldays, Ken recalls Kath Preston, who joined the staff of Burnley Grammar School in 1944. "We all fell in love with Katy Preston" said an old school friend. Ken recalls "She was in her early twenties, an outdoor girl ... genuinely interested in what she taught. She took us through 'Twelfth Night' for Higher School Certificate, and I have only to hear an allusion to that play and I am back again ... with our little group."

February 1960:

Kath

Long since, the partridges had called across the twilight fields,
And fallen silent. And cool and quiet lay the roadway home
That lovely evening half a lifetime gone.

I loved you so. Onto your shoulder set my hand,
Pretending to point out a star. You noticed, secretly.
So long ago; and yet we both remember.

Ken met Norma on a *British Trust for Conservation Volunteers* work party at Minsmere.

September 1962:

Norma

Haven
Blue sky and heather bride.

Star girl at evening
By the deep wood's edge
Whilst the nightjar sang
Below the last of sunset.

Dear hands holding:
Soft weight beautiful.

Autumn, winter; loneliness.
Sad beyond words, remembering the unforgettable;
Such love.

Seeking a sweetly-recollected dream,
I once again look back to lantern-light,
A full moon and the sea.

Ken and Barbara teamed up in 1964 – a partnership which was to last for forty years (Plate 29).

December 1964:

Barbara: at Nightfall

So quietly they went along the frosty path.
I might have been a man from any age,
Driving the sheep, at sunset, by the lake.

I held to heart the certainty
Of firelight; your gentle presence
And its sweet security.

A little loving and a winter fire:
Two blessings sought
By every man in every age.
Treasured
Against eternity.

Chapter 14
Days Out

In the course of a lifetime's bird-watching, extending over more than 60 years, it is surprising that there is one day which stands out. But such is the case. For Ken, 12th March 1957 was etched in his memory. It was a fine, spring day with blue skies on which he visited the countryside around Slaidburn with Arthur Welch. The hills were alive with the songs and courtship displays of waders – redshank, curlew, lapwing and snipe. On the same day Ken gives a graphic description of short-eared owls in flight. When one bird took cover in some birch trees, others twice tried to dislodge it, swooping in low with a crack as they clapped their wings, audible at about 150 yards. Others came together in the air and tilted before going in separate directions. There was a brief aerial clash with a carrion crow.

Ken taught me how to find lapwing nests. In one small field by the *Leeds and Liverpool Canal* above Altham, on 29th April 1959, I found ten nests. This was tantamount to colonial nesting and, except on land specifically managed for wading birds, is not a phenomenon one would come across now.

Another memorable day which I recorded in my natural history log was 7th May 1959 when I went with Ken and Eric Davis to the 'curlew country' around Gisburn and Paythorne. We caught the 6.55 a.m. bus from Burnley to Barnoldswick and, on alighting, walked towards Gisburn. We came to a field where Ken flushed a snipe and almost immediately found a snipe chick; it was crouching at the base of some rushes and was a rich chestnut brown with a few black markings and some silvery white, which provided good camouflage. The chick took fright, stumbling away on long legs and uttering high-pitched cheeps. Ken brought it back to where we found it and we left. Eric found a second chick a minute later. From a low-lying meadow, a curlew flew up silently, then began calling in a very distressed way; Ken found a nest lined with grass in which there were four eggs – bigger than a hen's and light green, with darker buff markings. Ken numbered them with an indelible pencil to see if the curlew shuffled them, but also to diminish their value to an egg-collector. At Gisburn was a field with three lapwings sitting on nests and one with three chicks. It had been a terrific field day.

In the same vicinity, Ken and I visited Hyles Moor above Bolton-by-Bowland on 6th July 1959 and counted a post-breeding flock of 120 curlews. On seeing us, they rose, trilling, circling on the wind to a great height, upon which a party of eighty flew westwards, whilst the other forty flew in a north-easterly direction - a sight not only of beauty, but one accompanied by their haunting calls so evocative of the northern moors.

The primary reason that these days in the field stand out in the memory now is that they seem like a distant dream. Fisher (1942) looked at future prospects for British birds. He predicted accurately that *"bird scientists will be able, with their armies of lay helpers, to conduct a great Mass Observation of birds"*, such as the *BTO's Atlas 2007-2011*. He foresaw some of the innovations which technology would bring: *"perhaps one day, not very far ahead, some zoologist may press a button in London or Oxford or Cambridge or Aberystwyth or Edinburgh, a machine will hum, and a stack of cards giving all the records of crossbill irruptions in East Anglia, or the history of pied flycatcher in Wales, or the migrations of the woodpigeon in Scotland, will fall into a tray."* Indeed, computation has now moved well-ahead of his vision. What Fisher, and indeed the whole of nature conservation, failed to foresee, however, was the sheer intensity of agricultural production after the war. He thought farmland bird populations would be healthy: *"the little birds of fields and hedges, nearly all friends of farmers and enemies of insects, will suffer less from human ignorance."* The fate of populations of farmland birds such as skylark *Alauda arvensis* and lapwing *Vanellus vanellus* would have shocked ornithologists such as Fisher and Oakes (1953). Gerry Almond told Ken that yellow wagtails *Motacilla flava* ceased to breed about 1996 when 'their field' by the River Calder at Altham was converted to silage. A letter of 17th March 2009 from Ken reads: *"I went on eastwards to see if skylarks or lapwings were back on the fields below Crown Point. None. I fear we are into The Silent Spring."*

In April 1960, I accompanied Ken and Derek Sodo to a gullery in a large amphitheatre in the Bowland Hills around the headwaters of Hare Syke and Gravells Clough, above Tarnbrook. From below it was hidden by the lie of the land, but we could hear the birds calling. From the highest point at Ward's Stone, we estimated there to be about 3,500+ gulls. The majority were lesser black-backs *Larus fuscus* with a sprinkling of herring gulls *L. argentatus* and a pair of greater black-backs *L. marinus*. The gulls stood in pairs around well-formed scrapes - deep cups, lined with grass, the bottom usually being peat but in which eggs had not yet been laid. The ground was littered with droppings, feathers and fish bones. Favoured preening places had been flattened.

A further phenomenon I experienced with Ken was visiting starling *Sturnus vulgaris* roosts and attempting to assess the number of birds using them. My natural history log describes such a visit on 6th July 1961, when he and I visited Dean Clough near Great Harwood. My notes record *"the actual flight-in began at 20.40 and was over by 21.00. Estimating in hundreds, as they flew in a fairly steady stream from far to the south of the plantation, we calculated 10,000 birds were using the roost. There wasn't the same ceremonial flight in from the fields as in winter."*

On 31st December 1966 Ken, Roger Hughes and I visited the RSPB's nature reserve at Leighton Moss. There, we teamed up with the warden, John Wilson, to clear willow scrub and plant alder saplings in their place. Thirty years later, Ken returned with Peter Hornby, to join John as he went about his morning's work, setting up nest-boxes for bearded tits. Ken records *"we went out in a punt from the causeway and away into the more secret creeks down the Moss. Absolutely superb. John and Peter poled the punt along; I just sat in the middle like the Prince Regent."* As for the alders we planted at the head of the Moss, these are now full grown and provide a food source for siskin *Carduelis spinus* and redpolls *C. flammea*.

On Christmas Day 2006, Sharon Parr and I joined Ken. Having left lunch cooking, we headed out to Sheldon Hushings between Cliviger and Towneley. Ken had been told of 200 or so bramblings *Fringilla montifringilla*, feeding there amongst beech trees. We parked in mist so thick we could only just make out one of the massive wind turbines a few yards away. We walked along an old pack-horse road, running gently downhill through a lunar landscape, rock-strewn in places, where pent-up hushing water had long ago been used to expose limestone in the glacial till. Groups of beech trees surmounted numerous hillocks of long-vegetated spoil, in one of which Ken spotted twenty bramblings; so murky was the light that, even within a few yards, we could only just make out their distinctive plumage. As we retraced our steps, a hen harrier wafted past, just visible through the mist.

In June 2007, I took Ken on an RSPB guided walk as part of the *Bowland Festival* to see the implementation of management sensitive to the needs of nesting waders. Gavin Thomas took us through fields which were being farmed under *Countryside Stewardship* at Lower Fairsnape. Several lapwing chicks, mostly so well-grown as to be almost fledged, were present. No machine operation is allowed during the breeding season and ditches must be gently profiled so that wader chicks can clamber out. Ken enjoyed the afternoon and was most impressed by what he had seen.

For his 80th birthday on 24th June 2008, I took Ken to Pendleton and we strolled the length of the village street. Swifts *Apus apus*, house martins *Delichon urbica* and swallows *Hirundo rustica* were all hawking for insects. We counted four swifts and Ken saw one enter and leave the eaves of a house, evidently breeding. House martins were the most numerous and we counted nine nests; three were in usual locations in the corners of windows or under eaves, but six were under the roof of a farm out-building where one would expect swallows' nests. We saw only two swallows. The householder with martins nesting under the eaves had evidently tried to dissuade them with wire-netting and hanging bamboo canes, but the martins had flown between the canes and used the wire-netting as a base for their mud-nests. With everything else these birds have to contend with, some householders are still reluctant to give them a temporary home.

References

Fisher, J. 1942. *The birds of Britain.* William Collins, London.

Oakes, C. 1953. *The birds of Lancashire.* Oliver & Boyd. Edinburgh and London.

Chapter 15
Butterfly Days

In 1992, Ken resumed a boyhood interest when he began compiling a local list of butterflies: *"I started with them before I went on to birds."* A series of visits to likely localities followed, often in the company of his long-time friend, the late Gerry Almond. On 2nd May 1994, during a visit to Longridge Fell, Gerry spotted a green hairstreak *Callophrys rubi* which was a lifetime first for them both. On a visit with Peter Hornby in May 2000 to the Langden Valley in Bowland, amongst the bilberry and heather, they saw many green hairstreaks. In May 2004, I was lucky enough to see them for myself when visiting an area of bilberry heath in the Cliviger Gorge which Ken thought a likely spot for them.

The rekindling of his interest also saw the start of an annual butterfly report, initiated by Ken, subsequently taken over by Jeff and Sheila Howarth, and later by John Plackett and Peter Hornby. This attracts records from other observers and in the long-term should be a good source of information for phenological purposes.

The first comma *Polygonia c-album* Ken ever saw in Lancashire was with Clifford Oakes and Arthur Welch on 16th May 1948 at Abbeystead in Bowland. It was not until a visit with Barbara to Spring Wood near Whalley on 29th April 1995, that he saw another, believed to be part of a northern extension of its range. By 1996, they were established at Spring Wood and at Martholme Viaduct and still spreading. In July 1996, he and Barbara saw a newly emerged comma at close range at Heasandford. This trend continued in 1997 and, in May 1999, Ken and Barbara even had one in their back garden at 167 Manchester Road and, in October 2001, one at Towneley Hall.

1995 proved to be the best year that Ken could remember for red admirals *Vanessa atalanta*: *"Did I tell you of the fantastic migration of them that*

I saw on Crown Point on 13th September? They were coming right past me, flying southwards at a rate of more than one per two minutes and, if I scanned the moor with binoculars, I could see three or four within the field of view quite easily."

In June 1996, Ken observed a painted lady *Cynthia cardui* over Crown Point and subsequently saw and heard of others – evidently part of an 'invasion' from Europe by this migratory butterfly on a scale he had never previously experienced. Their progeny began emerging in early August and Ken counted fifteen within a mile of home, mainly on buddleia and a few on creeping thistle *Cirsium arvense* until, by mid-August, he had seen scores of them. In April 2000, Ken and Gerry visited Barley and were reliably told of a painted lady. Ken commented *"the textbooks tend to say that it does not hibernate in England, but it certainly does sometimes."* There must have been a further invasion in 2003 because, in August of that year, there were *"quite a few painted ladies about ... all looking to be in mint condition."* In 2007, a spectacular arrival of red admirals and painted ladies in western Ireland was described by Michael Viney of the Irish Times; Ken commented *"we just caught the edge of this here, with a few painted ladies."*

April 1998 saw a few small tortoiseshells *Aglais urticae* on the wing and more peacock *Inachis io* butterflies than Ken had ever seen in spring before. These would have emerged from hibernation and he observed them nectaring on dandelions. He dubbed 2001 as 'The Year of the Peacock', having seen more than at any time previously. In September 2003, he saw literally hundreds of small tortoiseshells at Towneley.

On 4th April 1999, Ken was coming up the back garden path when Barbara appeared at the door and called *"It's for you"*. He supposed she meant a phone call and did not hurry unduly, whereupon she beckoned him urgently and called out again, when he realised she had said *"It's a blue"*. And so it was, a holly blue *Celastrina argiolus*, in the front garden. A second one was observed later in the spring.

Butterflies were scarce at first in 2000, but by August they had *"come into the scene at last"* and Ken had seen all the species that he would expect, plus a clouded yellow *Colias croceus* at Salthill, near Clitheroe.

Also in 2000, speckled woods *Pararge aegeria* were beginning to colonise the area around Burnley. Gerry told Ken of two at Cliviger in August 2001; he also saw a brimstone *Gonopteryx rhamni gravesi*, which is dependent on buckthorn, some of which had been planted in the Forest of Burnley, near Towneley, a couple of years previously. In September 2003, Ken commented on the fantastic increase in speckled woods with sightings at Healey Heights, Towneley, Scott Park and even at Clowbridge nearly 1000ft above-sea-level. Although they were only seen in ones and twos, the overall population probably numbered thousands. In July 2005, in a wooded glade near Habergham Parish Church, Ken showed me two speckled wood butterflies spiralling up in a characteristic dance until lost to view in the canopy.

New records continue to be made. In July 2005, he and Gerry saw a recent pioneer, the gatekeeper *Pyronia tithonus*, at Barley and, in July 2007, Jeff Howarth reported that several ringlets *Aphantopus hyperantus*, new to the area, had been seen at Rowley.

With Gerry Almond, a remarkable sighting at Ogden Lower Reservoir on 9th August 2009 was of a cluster of 31 green-veined whites *Pieris napi*. They were gathered on dark moss and sand within one square metre; Ken and Gerry surmised they might have been licking salts from fox or badger urine. Seeking to share what they had just witnessed, they described it to two walkers only to observe, to their surprise and dismay, that they did not seem interested and obviously did not realize the significance of this event.

Subsequent to these fulfilling butterfly days however, for reasons unknown, but possibly relating to the use of insecticides in increasingly intensive agriculture, or the impact of climate heating, numbers of all species have become very scarce.

Chapter 16
On life

Spencer's Law

The Guardian of 9th October 1967 contained a letter from Ken: *"The revival of the oystercatcher : cockle controversy affords a nice illustration of what I might lightheartedly call 'Spencer's Law' - that an increase in any avian species is immediately followed by a call for its reduction."*

John Robins of *Animal Concern*, writing to Ken in May 1994, said "I think ... 'Spencer's Law' is spot on – should any species show an increase in population the human influence on the situation is ignored and the species culled for its own good." He had just returned from a BBC Radio debate in Peterhead where over forty representatives of the fishing industry admitted that illegal landings of over-quota fish were commonplace. "Did they want grants to decommission their vessels or were they willing to be compensated to allow the North Sea to lie fallow for a few years? No; they wanted permission to cull 50,000 seals ..."

A letter to *The Times* of 22nd July 1996 from Rupert Ridge of Brockley Elm House, (a badger address!) was entitled *Badger Pests* and read *"A few days ago my wife and I found a badger sitting in our hen house surrounded by the gasping and bloody remains of our hens (not the first such attack) ... in their current numbers these magnificent animals are now also a pest."* Ken had annotated it *"Here we are again – another example of 'Spencer's Law', that any significant increase of a species is followed by a demand for its reduction. The letter has been answered by several correspondents in a subsequent edition, one of them making the obvious point that RR should have locked up his hen-pen better."*

Recently, following lunch at *The Swan with Two Necks* in Pendleton, Ken and I went for a stroll along a lane adjacent to Pendleton Hall. It was a sunny autumn day; from a wall festooned with flowering ivy came the hum of insects – bees, hoverflies and butterflies were partaking of the ivy's copious nectar. A year later, we sought to repeat the experience, only to find that the wall had been totally stripped bare. This led Ken to speculate whether he should not define 'Spencer's 2nd Law' along the lines that, *'no sooner have you discovered something to value and treasure, then in all likelihood someone will come along and destroy it'.*

Eyesight

In later years, Ken had to wear glasses and found this very frustrating, as recorded in the following verse:

> **25 Years a Presbyope**
> *I can't get used*
> *to getting used*
> *to not being able to see*
> *what I'm looking at.*

As the author of this biography, I too have now encountered eyesight problems. My eyes served me well until the age of 76, at which point developing glaucoma necessitated my ceasing to drive.

Soul Mates

Flora Britannica (1996) contains many references to John Clare. Ken identified closely with him. *"He was hardly heard of as a poet when I was at school but has rightly come into his own during my adult lifetime; an interesting man with whom I feel a great affinity."*

Having read George Mackay Brown's autobiography (1997), Ken saw himself *"as having a good deal in common with GMB, though not in his drinking or in his catholicism."*

Ken developed an interest in the Brontës: *"What a fantastically creative family! Strange how you can feel kinship sometimes with people long dead, more than with many contemporaries."*

A Radio 4 programme to which Ken made a point of listening was *Something Understood*. In an edition in November 2004, a guest referred to his father having been 'a miniaturist', that is, he focused on small and overlooked things, such as the first spring bulbs beginning to come through. Ken commented *"I realise that I am becoming more and more so, and am happier to let more major concerns (if such they are) pass me by."* Moss (2004) quoted James Fisher making the same point: *"Some people say that the popularity of natural history ... derives from the wish of many of our industrial population to escape; some even call it an escape from reality ... if it is an escape then it is **an escape to reality**."* (My emboldening.)

Ken's hero was undoubtedly Charles Darwin. He greatly enjoyed Rebecca Stott's *Darwin and the barnacles*: *"I see that Darwin was on the Isle of May on an early visit. So I have walked in the steps of a master, for as you know I was there for a week not long after the war."*

A biography Ken greatly enjoyed was that of *Thomas Hardy* by Clare Tomalin. He particularly liked *Afterwards*, in which to some extent he saw himself.

> **Afterwards**
> *When the Present has latched its postern behind my tremulous stay,*
>   *And the May month flaps its glad green leaves like wings,*
> *Delicate-filmed as new-spun silk, will the neighbours say,*
>   *"He was a man who used to notice such things?"*
>
> *If it be in the dusk when, like an eyelid's soundless blink,*
>   *The dewful-hawk comes crossing the shades to alight*
> *Upon the wind-warped upland thorn, a gazer may think,*
>   *"To him this must have been a familiar sight."*
>
> *If I pass some nocturnal blackness, mothy and warm,*
>   *When the hedgehog travels furtively over the lawn,*
> *One may say, "He strove that such innocent creatures should come to no harm,*

But he could do little for them; and now he is gone."
If, when hearing that I have been stilled at last, they stand at the door,
  Watching the full-starred heavens that winter sees,
Will this thought rise on those who will meet my face no more,
  "He was one who had an eye for such mysteries?"

And will any say when my bell of quittance is heard in the gloom,
  And a crossing breeze cuts a pause in its outrollings,
Till they rise again, as they were a new bell's boom,
  "He hears it not now, but used to notice such things?"

Assisted Dying

Both Ken and Barbara were members of the *Voluntary Euthanasia Society*, now known as *Dignity in Dying: your life, your choice*. In 2004, Barbara died a drawn-out and painful death from cancer of which Ken said *"no one should have to die as Barbara did."* She would have liked to have been able to bring her life to an end, on her terms, before the illness took away her pleasure in life.

In 2006, Lord Joffe promoted his *Assisted Dying Bill*. However, *The Times* of 13th May carried an article by Greg Hurst, political correspondent, entitled *"Peers wreck Bill to legalise euthanasia for terminally ill. A well-funded pressure group,* Care not Killing, *fought the Bill on behalf of churches, palliative care organisations and other groups ... Lord Joffe opened the debate by urging peers to recognise that the heavy mail bags they had received were generated by an unrepresentative lobby led by the Roman Catholic Church ... Prominent supporters of the Bill included Baroness Jay of Paddington ... who told peers 'We have to recognise that some terminally ill people would prefer to end their lives in a controlled and dignified manner, rather than receive care until a so-called natural death'."*

Ken read the main press reports and wrote to thank some of those who spoke in support of it and to reason with others who were against it. He considered that assisted dying must come, just as cremation had.

Professor Raymond Courteney Tallis, a specialist in geriatric care at the University of Manchester, was one of Kirsty Young's guests on Radio 4's *Desert Island Discs* in 2007. Ken wrote to him to say how much he had been heartened by his views on assisted dying. To Ken's surprise and delight Professor Tallis telephoned him and they had quite a long conversation about the issue.

In late 2007, Joan Bakewell hosted her series *Belief* on Radio 3 and one of her guests was Baroness Warnock. Ken considered her an intelligent and articulate person. It had been good to hear two people speak for half-an-hour without ever saying *"you know"*. Ken considered that her views on assisted dying would prevail in time.

In December 2007, Ken received a letter from Ashley Riley, Head of Campaigns and Communications, of *Dignity in Dying* regarding his correspondence with Lord Joffe. She wrote: *"I am sorry to hear of Barbara's death two years ago. You will not be surprised that many supporters of a change in the law are those people who have had to witness agonising deaths of loved ones ... We shall overcome."*

Ken took up the theme in May 2008 with Nicholas Reade, *The Lord Bishop of Blackburn*, recognising that while *"it is not easy for clergy to declare their support. My hope is that you may do so, should an opportunity arise."* The Bishop's Chaplain, the Reverend David Arnold, replied *"Bishop Nicholas is very strongly of the view that Assisted Dying is against the Law of God."* This was like a red rag to a bull. Ken responded *"I intend no disrespect when I now say that I regard the Bishop's views on the subject of Assisted Dying as medieval ...   My beloved Barbara ... would have liked the right to die when all hope of any recovery had gone."*

Badham (2009) published a book on the Christian case for assisted dying. Ken considered it brilliant and bought several copies, sending one to the Bishop of Blackburn and another to Joan Bakewell.

Ken missed Barbara terribly and recalled his friend Clifford Oakes quoting lines, in reference to his wife Annie, from Tennyson's *Break Break*:

> *But O for the touch of a vanish'd hand,*
> *And the sound of a voice that is still!*

A second poem Ken quoted from is *Absence* by Mary Sarton.

> *Must we lose what we love*
> *To know how much we loved it?*

> *It is always there now,*
> *That absence, that awful absence.*

Recipe for Happiness

Ken's recipe for happiness: 'someone to love, something to do and something to look forward to'.

References

Badham, P. 2009. *Is there a Christian case for assisted dying? Voluntary euthanasia reassessed*. SPCK.

Brown, G. M. 1997. *For the islands I sing: an autobiography*. John Murray, Edinburgh.

Mabey, R. 1996. *Flora Britannica*. Random House.

Moss, S. 2004. *A bird in the bush: a social history of birdwatching*. Aurum Press. p.183.

Stott, R. 2008. *Darwin and the barnacle*. Faber and Faber.

Tomalin, C. 2007. *Thomas Hardy*.

Chapter 17
Recognition
Plate 31

When Barbara was dying, she insisted that Ken – with all his many interests – still had much to live for. For 25 years, Karen Strachan had come in once a week to clean for Barbara and Ken; Barbara asked her to continue coming after her death. Karen would arrive just after breakfast on a Wednesday and make a cup of tea for them both. Ken looked forward to their weekly chats. He would then take himself out of her way by going down to work as a volunteer in Burnley Central Library.

Another landmark in his week was provided when, on Sunday mornings, his long-time friend the late Gerry Almond came round; together they would go out in Gerry's car to a variety of haunts seeking this bird or that butterfly.

After Barbara died, I began to visit Ken on his birthday and at Christmas. His birthday fell in mid-summer; we would sally-forth over the Nick-o'-Pendle for lunch at *The Swan with Two Necks* in Pendleton or The Lower Buck in Waddington. A distinguishing feature of both these country pubs was an absence of 'piped' music; Ken's hearing was deteriorating and extraneous noise would have made conversation impossible. At *The Swan* Ken was fascinated by a collection of poultry beside the beer garden.

The first Christmas after Barbara's death was more challenging. I knew Ken well enough to know that if, in conversation, I suggested coming to prepare Christmas meals for him, his spur of the moment response would be 'no'. So, I put the proposal in a letter and, sure enough, he told me later that his inclination had indeed been to decline. But then he thought: *"What would Barbara have said?"* and knew she would have said *'Go for it'*. And so the tradition was established of going out for lunch on the 23rd followed by a trip to Booth's in Clitheroe (where my mother formerly shopped) to stock up. On Christmas Eve, Ken always attended the afternoon carol service in *St Peter's Parish Church*, while Sharon and I prepared an evening meal and began preparing Christmas lunch. On the day itself, we would go for a morning walk, home for Christmas lunch and then retire to open gifts – those for Ken being almost invariably books.

The Weavers' Triangle

The two major events to sustain Ken throughout the year were his daily visits to assist in the Library and, from Easter to October, guiding visitors at the *Weavers' Triangle*. At both of these venues Ken was part of a family. Ken credited Brian Hall and Roger Frost, both former history teachers, and the late David Wild (1931-2014) – a local artist, with being the driving force behind the visitor centre, but his 'family' here included Roger and Dorothy Creegan, Lawrence Chew, the late Lynn Millard (1926-2013), the one-time Reference Librarian Jean Siddall and Mick and Diane Malone also formerly on the staff of the Library. Ken thoroughly enjoyed meeting visitors, some retracing their Burnley roots, and would help them with their queries about the town or its people, particularly their ancestors.

Burnley Central Library

Ken took a lifelong interest in libraries, commencing with a post with *Leeds City Libraries* in 1954. As his interest in community history intensified, so did his affiliation with libraries. Reference libraries provide the source material for history research and are the natural repository of findings. His wealth of experience spanned 60 years, the last 40 as a volunteer. This led to Ken receiving a huge surprise. He was nominated by the library staff as Lancashire's *Volunteer of the Year*. Ken was deeply touched by the nomination which read *"Ken, we wanted to show our enormous appreciation for all you do for us and the Library Service and give you an insight into why we care about you so much, Love from all your friends at Burnley Library."*

Among the dedications:

- Cath Atkinson – *"all staff recognize him as very much a part of our library family and a very valuable resource himself"*;
- John Davis – *"Ken has taught me to look at my home town from a different perspective"*;
- Neil Marshall – *"Ken is 'Mr. Community History"*;
- Catherine Fenton – *"so approachable and always helpful and kind"*;
- Susan and the team at Coal Clough – *"his knowledge knows no bounds"*;
- Shibli Begum – *"when we are completely stumped – you can bet Ken has either written something about this or he knows where he can find a small nugget of information"*;
- Sandra McGowan – *"his knowledge and enthusiasm about Burnley town and countryside is amazing"*;
- Rebecca Hewitt – *"his knowledge of the information held within the Community History Library at Burnley is second to none"*;
- Alison Hey – *"he has always considered the staff as his family and treats us regularly to chocolate."*
- Susan Broderick – *"I know I can ask for his help if I need to answer queries"*;
- Carole Wolstenholme – *"to me Mr. Spencer represents the true meaning of service; to give without any thought of reward; whilst always exuding the grace and good manners of a bygone era"*;
- Julie Dawson – *"he is one of the most interesting, inspiring, humble, kind and unassuming gentlemen I have ever met"*;
- Karl Munslow – *"he knows Walter Bennett's "History of Burnley" like the back of his hand. He can usually help us with the street and place

names, histories of notable local people and families … Google and Wikipedia mean nothing to Ken; index boxes, cards, microfilm, fiche and his memory of a lifetime's reading are his methods of solving things."

- Beverley Burrows – "Whenever Ken has been asked to help with a query he will do a tremendous amount of research and won't give up until he finds the answer".

- Anne Criscenzo-Whittam – "Whilst over the years staff have come and gone, Ken has been a constant, always enthusiastic to share his lifetime of knowledge and expertise with grateful staff and public alike."

- Mary Brown, District Manager, Burnley Libraries – "Ken is an inspiration to us all. Unfailingly polite and helpful, a true gentleman."

- David Johnston-Smith – "Born and bred in Burnley, he has a huge wealth of knowledge and experience in both the local and natural history of the town and its surroundings, and has published extensively about both subjects over the decades."

- Jane Morris – "I have worked in several roles at Burnley Library and whatever post I was in, I have always found him helpful and supportive."

- Catherine Park – "Every week, whatever the weather, Ken is at the library to help us. He is remarkable, walking down and up Manchester Road, come what may. Ken is unique as a volunteer and as a person, a true gentleman (very rare) and to sum up – "We shall not see his like again""

With this endorsement of the high esteem in which he was held as a volunteer, Ken attended the first round of judging for the East Lancashire Pride Awards which was held at Helmshore Museum, Rossendale. He travelled there with Julie Dawson and, although he regarded modern technology askance, he was nevertheless amazed at the satnav in her car, with its instructions on which turnings to take. The outcome of the evening was that Ken won the East Lancashire area award.

Ken's name then went through to the Lancashire County Council finals, to be held at Barton Grange Hotel on 11th September 2014 which was to be a formal awards dinner with entertainment provided by the Lancashire Youth Jazz Orchestra. This was not Ken's scene at all; rather than attend he was represented by library colleagues and, in his absence, awarded the overall title of **Volunteer of the Year**.

At the request of colleagues, as a precaution, Ken had written an acceptance speech which, in the event, was read out on his behalf:

"Recipients of prizes like the Oscars often say that they were astonished, which is sometimes hard to believe. In my case with the Pride Award it is absolutely true. I had never seen myself as a volunteer, but simply as a local historian helping readers and staff wherever I could.

I have been fortunate in my grandparents, my parents, my brother Philip (1922-1943) and my partner Barbara (1922-2004).

I am fortunate now with my friends. I treasure the book of testimonials compiled for my nomination. It means more to me than a Royal Honour."

There was more than a touch of irony in his reference to a Royal Honour. it was not until he was going through Barbara's papers after her death, that he discovered that she had put his name forward as a candidate for such an honour. But his involvement in the Burnley by-laws controversy had put paid to that. Recalling Brian Leveson QC's characterization of the *Burnley Dog Owners' Action Committee* (see Chapter 8), of which Ken had been a member, as *"the kind of people against whom Society needs to be defended"*, Ken chortled at such a strong body of proof to the contrary. He even wrote to the now ennobled Lord Leveson, to point out the error of his judgment.

Member of the British Empire

A member of *Burnley Central Library* staff, Julie Dawson, determined to approach *The Cabinet Office*, to nominate Ken for an Honour. In this, she was supported by Susan Halstead – formerly Research Librarian, and myself. Julie's efforts paid off and, in due course, Ken received a letter asking whether, if nominated, he would be prepared to accept an honour. With typical modesty, Ken pondered this offer for quite some time before deciding to accept. His reason for hesitating revolved around thoughts of Barbara. She would have been so proud of Ken and so delighted; but it seemed cruel that her earlier attempts had been thwarted. In due course, knowing that Barbara would have wanted him to accept the honour, Ken decided to do so.

And so, on 13th June 2016, at the age of 87, Ken was duly presented with *The Award of Member of the Order of The British Empire* by T*he Vice Lord-Lieutenant of Lancashire, Colonel Alan Jolley OBE,* in the presence of Mrs. Mary Jolly, Julie Bell – Head of Libraries for Lancashire County Council, Marian Taylor – Manager of Burnley and Pendle District Libraries, the Mayor Jeff Sumner and Mayoress of Burnley and invited guests.

Citation read at the presentation ceremony, 13th June 2016, in Burnley Central Library by Mr. Terry Hephrun:

MR KENRICK PARKER GRANT SPENCER

FOR THE AWARD OF

MEMBER OF THE ORDER OF THE BRITISH EMPIRE

Mr Spencer has volunteered at Burnley Library for over four decades, using his incredible knowledge to benefit thousands of people in the town. For more than 40 years, he has given over 15 hours every week volunteering in the Community History Department, helping both staff and users with their enquiries and their research. He has unrivalled expertise on the history of Burnley, which he has shared with local people for many years. He promotes the library's ethos of free access of information. Beyond answering queries from the public, he has developed the library's invaluable collection of local history resources by the publication and donation of his own books, pamphlets, photographs and leaflets. He has personally helped to develop these resources by the publication of over 200 authoritative books, pamphlets and leaflets with original research on the history of Burnley. In particular, his photographic collection of Burnley over the years ensures that the changing face of the local landscape has been comprehensively documented so people will have a true picture of Burnley, past and present. In 2014 at the age of 84, he was awarded Volunteer of the Year at the Lancashire County Pride Awards final.

Alongside his dedicated service to Burnley Library, he has also contributed to other local initiatives. He was a founding member of Burnley and District Civic Trust and is still actively involved in the campaigning work the group does to protect the built and natural environment in Burnley. He volunteers as a guide in the Weavers' Triangle Visitor Centre, where he imparts his knowledge to visitors on Burnley's industrial past. He has a keen interest in the environment and wildlife of the area, and for over 50 years has been involved in maintaining the local landscape through removing invasive shrubs and teaching local scouts and guides about the importance of protecting it. He is a keen ornithologist, even writing an authoritative book on the lapwing and promotes the study of birds throughout the area. At the age of 87, he continues to give his time freely and is an inspiration to many people in the area.

KEN'S RESPONSE

I can only repeat what I said on winning a County Pride Award as a Volunteer a couple of years ago. That is, whatever I may have achieved is due to my grandparents, my parents, my brother Philip (1922-1943), and to my partner Barbara Bailey (1922-2004). I have been very fortunate in my family and friends.

Afterwards, Ken was able to mingle with guests and enjoy a reception kindly provided by the Library staff. It was a very happy occasion; Ken subsequently commented on how lucky he was to be surrounded by such love and affection.

Annex

Letter to British Trust for Ornithology

24th August 2015

To:
Dr Andy Clements
Director, BTO
The Nunnery
THETFORD
Norfolk
IP24 2PU

From:
KG Spencer
167 Manchester Road
BURNLEY
Lancashire
BB11 4HR

Dear Dr Clements

BTO Gold Pin

Well, this was a surprise! Thank you all. I don't know how much you know of my involvement with the Trust. Here is an outline for you.

My acquaintance with it began in about 1934, when my brother Philip (who was killed in the war) started us on a Garden Bird Census at home. I'm sure he got the idea from Zoo magazine, of which James Fisher was the editor and, of course, as you know, he was a founder of the Trust. I was to meet him later, in about 1944, in Manchester, when he initiated the Rook investigation. Clifford Oakes was in charge locally, but he was working that afternoon – a Saturday – and asked me to go in his place. It was he who had introduced me to the Trust just a few weeks before.

I used to attend the conferences at Swanwick, and in one year, I joined with John Wilson of Leighton Moss to give a presentation on Lancashire ornithology.

I was the Trust's first Regional Representative for the county's eastern section, and remained so until I made way for Tony Cooper in about 1978.

I was the regional organizer for the 1968-72 Breeding Bird Atlas.

I have had correspondence with several members of the staff quite recently, on the question of bird populations being affected adversely by massive insect declines caused, I believe, by an increase of air pollution. Dr Stephen Ward of Arnside has helped me in promoting this supposition, as I have no computer. I enclose for you a copy of a letter from him to Dr Harrington at Rothampsted Research Station, written earlier this month.

I'll be writing to Stephen now, of course. I'm sure he will share my pleasure about the Gold Pin.

Thank you for your good wishes

Yours sincerely

Ken Spencer